Analysis of Machine Elements
Using COSMOSWorks

COSMOSWorks 2008

John R. Steffen, Ph.D., P.E.
Valparaiso University

ISBN 978-1-58503-474-1

Schroff Development Corporation
www.schroff.com

Trademarks and Disclaimer

SolidWorks and its family of products are registered trademarks of SolidWorks Corporation a Dassault Systemes, S.A. company. COSMOSWorks is a registered trademark of Structural Research & Analysis Corporation. Microsoft Windows and its family of products are registered trademarks of the Microsoft Corporation.

Every effort has been made to provide an accurate, error free text. The author and manufacturers shall not be held liable for any parts developed or designed based on this book or for errors appearing in this text.

Copyright © 2008 by John R. Steffen

All rights reserved. This document may not be copied, photocopied, reproduced, transmitted, or translated in any form or for any purpose without the express written consent of the publisher, Schroff Development Corporation.

About the Cover

Two-Stage Compressor, pencil on hard-board by Michael R. Steffen

An artistic interpretation of a mechanical device is chosen for the cover because, although finite element analysis is a highly sophisticated, mathematically based software program, its application requires a certain amount of art to be used properly.

Examination Copies

Books received as examination copies are for review purposes only and may not be made available for student use. Resale of examination copies is prohibited.

Electronic Files

Any electronic files associated with this book are licensed to the original user only. These files may not be transferred to any other party.

Dedication

For family and friends who have provided guidance, inspiration, encouragement and much delight. Especially my wife Marolyn, and children Amy and Michael.

About the Author

Dr. John Steffen obtained his B.S.M.E. from Valparaiso University, M.S.M.E. from The University of Notre Dane, and Ph.D. from Rutgers, The State University of New Jersey. His areas of interest include Machine Design, Mechanism Synthesis, Finite Element Analysis, and Experimental Stress Analysis. Dr. Steffen is Professor of Mechanical Engineering and occupies the Alfred W. Sieving Chair in Engineering at Valparaiso University, Valparaiso IN. He is an active member of the American Society of Mechanical Engineers (ASME) and the American Society for Engineering Education (ASEE).

Acknowledgements

The author would like to acknowledge the following individuals and corporations for their support, encouragement, and invaluable assistance in preparation of this book.

- The SolidWorks Corporation for their support and assistance.

- Mr. Stephen Schroff and Ms. Mary Schmidt of Schroff Development Corporation, for their guidance and assistance preparing the manuscript for publication.

- Mr. Michael A. Steffen, for his assistance with preparation of many SolidWorks models and drawings.

- Ms. Cami Gudino, Administrative Assistant in the College of Engineering, for text entry and knowing the answers to numerous logistical questions.

- Mr. Joe Galliera, SolidWorks Corp., and Mr. Ross Mack, for troubleshooting assistance within both SolidWorks and COSMOSWorks.

- Valparaiso University and the College of Engineering who provided an environment which encouraged and supported this effort.

- Magdalen Fryatt and the SolidWorks Solution Partner team.

This publication marks the 3nd edition of a learner centered, step-by-step guide to using COSMOSWorks. Every attempt has been made to make this book error free so that it best serves first-time COSMOSWorks users. In many instances, the software provides alternative approaches to perform the same tasks. Some alternatives are presented, however, it is not possible to present every option while keeping the text of manageable length. In an effort to improve future updates of this text, the author welcomes error corrections and suggestions to improve the presentation at john.steffen@valpo.edu.

NOTES:

Table of Contents

Table of Contents — i

Preface — vii

Intended Audience for this Text — vii
Using this COSMOSWorks User Guide — vii
Instructors Preface — ix

Introduction

Finite Element Analysis — I-1
Nodes, Elements, Degrees of Freedom, and Equations — I-2
COSMOSWorks Elements — I-3
Solid Elements — I-3
Solid Element Degrees of Freedom — I-4
Shell Elements — I-4
Shell Elements Degrees of Freedom — I-5
Meshing a Model — I-5
Introduction to the COSMOSWorks User Interface — I-7
    Orientation and Set-up of SolidWorks Work Environment — I-7
    Orientation to the COSMOSWorks Work Environment — I-11
    Property Managers and Dialogue Boxes — I-12

Chapter 1
Basic Stress Analysis Using COSMOSWorks — 1-1

Learning Objectives — 1-1
Problem Statement — 1-1
Creating a Static Stress Analysis Study — 1-3
Assigning Material to the Model — 1-4
Applying Restraints — 1-6
Applying Loads — 1-8
Meshing the Model — 1-11
Running the Solution — 1-13
Examination of Results — 1-14
Default COSMOSWorks Graphical Results — 1-14
Results Predicted by Classical Stress Equations — 1-15
COSMOSWorks Results for Stress in Y-Direction — 1-16
Using the Probe Tool — 1-18
Summary — 1-25
Exercises — 1-26

Chapter 2
Curved Beam Analysis   2-1

Learning Objectives	2-1
Problem Statement	2-1
Creating a Static Analysis (Study)	2-2
Assign Material Properties to the Model	2-4
Applying Restraints	2-6
Applying Load(s)	2-7
Inserting Split Lines	2-8
Applying Force to an Area Bounded by Split Lines	2-11
Meshing the Model	2-12
Solution	2-14
Examination of Results	2-15
Analysis of von Mises Stresses Within the Model	2-14
Verification of Results	2-18
Results Predicted by Classical Stress Equations	2-18
Comparison with Finite Element Results	2-20
Assessing Safety Factor for the Curved Beam	2-23
Reaction Forces	2-28
Logging Out of the Current Analysis	2-29
Exercises	2-30

Chapter 3
Stress Concentration Analysis   3-1

Learning Objectives	3-1
Problem Statement	3-1
Create a Static Analysis (Study)	3-2
Defeaturing the Model	3-3
Assign Material Properties to the Model	3-4
Apply Restraints and Loads	3-5
Meshing the Model	3-6
Solution	3-8
Examination of Results	3-8
Stress Plots	3-8
Creating a Copy of a Plot	3-11
Displacement Plot	3-13
Creating New Studies	3-15
Study Using a High Quality Elements and COARSE Mesh Size	3-15
Study Using a High Quality Elements and DEFAULT Mesh Size	3-20
Study Using a High Quality Elements and FINE Mesh Size	3-21
Study Using High Quality Elements and Mesh CONTROL	3-21
Results Analysis	3-26
Create Multiple Viewports	3-26
What Can Be Learned from this Example?	3-28

Table of Contents

Other Uses of the Copy Feature	3-28
Comparison of Classical and FEA Results	3-32
Exercises	3-33

Chapter 4
Thin and Thick Wall Pressure Vessels

Learning Objectives	4-1
Thin-Wall Pressure Vessel	4-1
Problem Statement	4-1
Defining Options at Start of a Study	4-4
Creating a Static Analysis Using Shell Elements	4-10
Assigning Material Properties	4-10
Assigning Loads and Restraints	4-11
Symmetry Restraints Applied	4-11
Pressure Load Applied	4-14
Meshing the Model	4-14
Solution	4-15
Results Analysis	4-16
Thick Wall Pressure Vessel	4-21
Problem Statement	4-21
Defining the Study	4-21
Assign Material Properties	4-22
Define Restraints and Loads	4-22
Mesh the Model	4-25
Solution	4-27
Results Analysis	4-27
Displacement Analysis	4-27
Von Mises Stress Analysis	4-29
Tangential Stress Analysis	4-29
Adjusting Stress Magnitude Display Parameters	4-31
Using Section Clipping to Observe Stress Results	4-34
Exercises	4-38

Chapter 5
Interference Fit Analysis — 5-1

Learning Objectives	5-1
Problem Statement	5-1
Interference Check	5-2
Creating a Static Analysis (Study)	5-3
Assign Material Properties to the Model	5-4
Defeature the Model	5-5
Apply Loads and Restraints	5-6

Un-suppress Part of the Model to Use Symmetry	5-6
Define Symmetry Restraints	5-7
Apply Restraints to Eliminate Rigid Body Motion	5-8
Define a Shrink Fit	5-10
Mesh the Model and Run the Solution	5-13
Examination of Results	5-14
Default Stress Plot	5-14
Stress Plots in the Cylindrical Coordinate System	5-17
Circumferential (Tangential or Hoop) Stress	5-17
Radial Stress	5-20
Verification of Results	5-22
Stress Predicted by Classical Interference Fit Equations	5-22
Stress Predicted by Finite Element Analysis	5-23
Radial Stress Comparison	5-23
Circumferential Stress Comparison	5-25
Quantifying Radial Displacements	5-26
Generating a Report	5-28
Exercises	5-31

Chapter 6
Contact Analysis in a Trunion Mount

Learning Objectives	6-1
Problem Statement	6-1
Preparing the Model for Analysis	6-2
Add Reference Planes	6-3
Insert Split Lines	6-4
Creating the Assembly Model	6-5
Create a Finite Element Analysis (Study)	6-10
Assign Material Properties	6-10
Cut Model on Symmetry Plane	6-10
Assign Restraints and Loads	6-14
Symmetry and Immovable Restraints	6-14
Contact/Gaps Restraints	6-14
Apply a Directional Load to the Pin	6-16
Meshing the Model and Running the Solution	6-19
Results Analysis	6-19
Von Mises Stress	6-19
ISO Clipping	6-20
Animating Stress Results	6-22
Displacement Results	6-23
Contact Pressure/Stress	6-24
Exercises	6-26

Chapter 7
Bolted Joint Analysis

Learning Objectives	7-1
Problem Statement	7-1
Create a Static Analysis	7-2
Assign Material Properties to the Model	7-2
Apply Loads and Restraints	7-3
Traditional Loads and Restraints	7-3
Define Bolted Joint Restraints	7-4
Define Local Contact Conditions	7-11
Mesh the Model and Run Solution	7-13
Results Analysis (Downward Load)	7-14
Von Mises Stress	7-14
Bolt Forces	7-15
Define a New Study with the Applied Load Reversed	7-16
Results Analysis (Upward Load)	7-17
Von Mises Stress	7-17
Bolt Forces (Upward Load)	7-17
Bolt Clamping Pressure	7-22
Summary	7-26
Exercises	7-27

NOTES:

PREFACE

Intended Audience for This Text

This text is written primarily for first-time COSMOSWorks® users who wish to understand finite element analysis capabilities applicable to stress analysis of mechanical elements. The focus of examples is on problems commonly found in an introductory, undergraduate, Design of Machine Elements or similarly named courses. Also, depending upon the primary textbook used for a Mechanics of Materials course, many examples contained herein can also be introduced in those courses. However, those courses are usually so filled with new concepts that adding finite element analysis to the mix of topics only serves to dilute the understanding of both subjects. Other novice users of COSMOSWorks, possessing the background described above, will also benefit by using the self-study format used throughout this text.

Many chapter example problems are accompanied by problem solutions based on use of classical equations for stress determination. This approach amplifies a fundamental tenet of this text, which is that a better understanding of course topics related to stress determination in mechanical elements is realized when classical methods *and* finite element solutions are considered together.

Although it is assumed that readers have working knowledge of SolidWorks, practicality dictates that many users will be new to the SolidWorks/COSMOSWorks work environment or transitioning to it from other finite element software programs. For these reasons, this text is organized so that individuals with no prior SolidWorks or COSMOSWorks experience will be successful. To assist overcoming any lack of SolidWorks familiarity, models of parts or assemblies for all examples and end of chapter problems can be downloaded from: http://www.schroff1.com/.

Using this COSMOSWorks User Guide

Each chapter of this text begins with a list of *Learning Objectives* related to specific capabilities of the COSMOSWorks program introduced in that chapter. Most software capabilities are repeated in subsequent examples so that users become familiar with their purpose and are capable of applying them to future problems. However, successive use of repeated steps is typically accompanied by briefer explanations. This approach is used to minimize document length, to permit users to work more at their own pace, and to devote more emphasis to new concepts being introduced.

Unlike many step-by-step user guides that only list a succession of steps, which if followed correctly lead to successful solution of a problem, this text attempts to provide insight into *why* each step is performed in a "just-in-time" manner. A consequence of this approach is somewhat lengthier explanations of new topics. However, these explanations ensure a deeper understanding and ultimately enhance knowledge about the software and the finite element method.

Numerous traditional design of machine elements textbooks[1,2,3,4,5,6] were reviewed prior to embarking on the writing of this COSMOSWorks user guide. The goal was to identify common organizational schemes such that this text would be compatible with the bulk of commonly used undergraduate machine design texts. However, considerable organizational differences between textbooks, compounded by the fact that course outlines often vary by instructor, made it unlikely that a common format would accommodate all organizational needs. Perhaps the only common theme found is that chapters in the first-half of the referenced texts focus on topics related to special states of stress found in mechanical elements.

For these reasons, this text begins with problems that can be solved with a basic understanding of mechanics of materials. Problem types quickly migrate to include states of stress found in more specialized situations common to a design of mechanical elements course. Paralleling this progression of problem types, each chapter introduces new software concepts and capabilities. Therefore, it is recommended that chapters be followed in the sequence presented. However, despite this recommendation, each chapter is self-contained, thus examples can be worked in any order.

Each example is divided into several major sections that are designated by a descriptive title or a sub-title. This is done to focus attention upon a specific overall task within each section. For example, individual sections are devoted to: material selection, application of restraints and applied loads, meshing the model, etc. Each section is subdivided into a series of numbered steps intended to lead the user through a logical sequence of actions necessary to accomplish the overall goal of a specific section. Numbed steps do not necessarily imply a rigid order to an analysis. Rather, the numbers are to serve as "location finders" when working back-and-forth between this user guide and a computer screen. Each menu selection is printed in **bold font** to facilitate locating the exact word or phrase on the computer monitor.

Because COSMOSWorks is a Windows® based program, the left mouse button serves the usual purposes of clicking to select an item or clicking-and-dragging to draw a line and/or to create or select a geometric shape. References to left mouse button selections are simply denoted by the words "click" or "select." Clicking the right mouse button provides access to numerous pull-down menus within COSMOSWorks. A "right-click" is always specifically indicated where applicable. Selections from within pull-down menus are always made with the left mouse button.

---

[1] Collins, J.A., <u>Mechanical Design of Machine Elements and Machines</u>, John Wiley & Sons, Inc, 2003.
[2] Hamrock, B., Schmid, S.R., Jacobson, B., <u>Fundamental of Machine Elements</u>, 2nded., McGraw-Hill, 2005.
[3] Mott, R.L., <u>Machine Elements in Mechanical Design</u>, 4th ed., Pearson-Prentice Hall, 2004.
[4] Budynas, R.G., Nisbett, J.K., <u>Shigley's Mechanical Engineering Design</u>, 8thed., McGraw-Hill, 2008.
[5] Spotts, M., Shoup, T.E., Hornberger, L. <u>Design of Machine Elements</u>, 8thed., Pearson-Prentice Hall, 2004.
[6] Ugural, A.C., <u>Mechanical Design an Integrated Approach</u>, McGraw-Hill, 2004.

It is frequently necessary to reorient COSMOSWorks solid models or assemblies by zooming, rotating, or otherwise manipulating the model to facilitate application of loads or restraints to specific geometric entities. To facilitate these operations without the need for menu or icon selections, the following "short-cut" key and mouse combinations are summarized in Table 1.

Table1 - Short-cut key and mouse button selections

Key and Mouse Button Combinations	Resulting Model Motion
Roll the Middle Mouse Button (MMB) 'up' or 'down.'	Zoom 'out' or 'in' on the model, respectively, in *incremental* steps.
[Shift] + MMB and push mouse 'away' from or 'toward' the user.	Zoom 'in' or 'out' on the model respectively, by *smooth* motion.
Press MMB and move mouse.	Rotates the model rotates in 3-dimensions.
[Ctrl] + MMB and move mouse.	Model moves left or right and/or up or down on the graphics screen.

Instructors Preface

This text is intended for use by students with background in an introductory Mechanics of Materials course. However, the focus of most examples and chapter problems is on topics commonly found in a Design of Machine Elements or similarly named course.

While it is fairly common knowledge that virtually anyone with a technical background can become proficient at using a finite element program, what is frequently lacking is the ability to discern meaningful results from the copious output produced by such programs. To address this perceived weakness, this text consistently attempts to compare finite element results to results found using classical equations of stress analysis.

To accommodate organizational differences between design of machine elements textbooks and between individual course outlines, chapters of this text can be worked in any order. However, it is strongly recommended that Chapters 1 and 2 serve as a common starting point. While subsequent chapters may refer to techniques mastered in earlier (skipped) chapters, all necessary analysis steps are included so that users are able to complete each example. Also, due to default software output that uses von Mises stress plots, this topic is briefly introduced in Chapter 2. Instructors should be aware of this fact and attempt to fill in any gaps in understanding until von Mises stress is addressed in your course.

Finally, "check sheets" are provided to facilitate grading all end-of-chapter problems. These check sheets are created in MS Word® and, thus, can easily be edited to emphasize (add or delete) particular aspects of a problem and/or to change point values associated with portions of an analysis. Check sheets can be downloaded at the publisher's web site: http://www.schroff1.com. This is the same web site where model files for all textbook problems are found. It can be instructional to provide these sheets to students when problems are assigned so that expectations are clearly understood.

NOTES:

INTRODUCTION

Finite Element Analysis

Finite element theory was introduced more than 60 years ago. However, implementation of this theory only became practical with the advent of high-speed computers. This section introduces one method of understanding the numerous sets of simultaneous equations that must be solved as part of the finite element method, and therefore, why this technique requires computer solution. Other more refined mathematical formulations exist, however, they are not the subject of this text. Suffice it to say that finite element theory applied to the solution of any realistic problem results in a computationally intensive exercise. Additional insight into the number of equations that must be solved in a finite element analysis is related to the number of nodes and elements in a model as described in the next section.

The following discussion provides a simplified overview of the mathematical basis of a finite element solution based on the *stiffness* approach. Begin by considering a simple member of original length **L** subject to an external axial load **F** as shown in Fig. 1. Due to force **F**, the member undergoes an axial deformation shown as Δ**L**. For this simple case, the well-known equation relating force and deformation is given by equation [1].

$$\Delta L = \frac{FL}{AE} \qquad [1]$$

Notice that the material property (**E** = modulus of elasticity) along with the applied force **F**, length **L**, and cross-sectional area **A**, must be known to solve for Δ**L**.

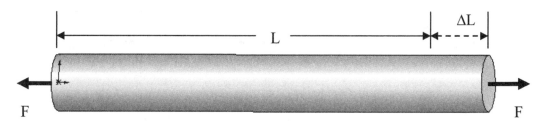

Figure 1 – A simple member subject axial force **F** to illustrate the relationship that exists between the applied force and the resulting deformation Δ**L**.

The next step in a stiffness formulation is to compute strain based on the deformation that was determined using equation [1] above. To do so the simple definition of strain is used; it can be expressed verbally and mathematically as:

strain = (change of length) / (original length)

or

$$\varepsilon = \frac{\Delta L}{L} \qquad [2]$$

Finally, since the goal of analysis is determination of stress in a member, stress can be can be determined from the classic stress-strain relationship listed below.

$$\sigma = E\varepsilon \qquad [3]$$

Of course, equation [3] is valid only in the elastic region where stress is proportional to strain. This fact is fundamental to *linear* finite element analysis described in this text.

Thus, finite element analysis begins with a basic mathematical description of deformation and proceeds to the solution of member stresses. Other formulations of finite element equations exist, however, they are not described here.

Nodes, Elements, Degrees of Freedom, and Equations

To permit mathematical analysis by the finite element method requires simplification of modeled parts. If the cylindrical member, shown in Fig. 1, were divided into an arbitrary number of *nodes* and *elements*, its model might appear as shown in Fig. 2.

Figure 2 – Simplified model of the cylindrical member sub-divided into a series of nodes and elements.

In Fig. 2, *elements* are represented by short line segments between successive numbered points. The small numbered circles, which represent points of connection between adjacent elements, are called *nodes*. For tracking purposes, elements are also numbered within finite element software. However, for simplicity, they are not numbered in Fig. 2.

Many different types of elements are found in commercial finite element software. COSMOSWorks beam elements can be used to model the part shown in Figs. 1, 2 and 3. Beam elements are included here to introduce the concepts of *degrees of freedom* and the *number of equations* that result during a finite element solution. To assist with this understanding, we isolate an arbitrary element and its two nodes (n and n+1) from Fig. 2 to obtain an enlarged view of a single element and its two nodes shown in Fig. 3 below.

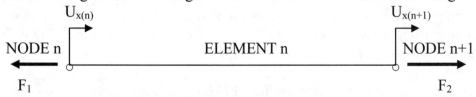

Figure 3 – A simple one-dimensional model represented using nodes and an element.

First, we assume that displacements at nodes n and n+1 are restricted to lie in the X-direction. This restriction ensures that the problem is one-dimensional (i.e., displacements are limited to a single direction, the X-direction). $U_{x(n)}$ and $U_{x(n+1)}$ represent possible displacements in the X-direction at *each* node. These displacements correspond to one *degree of freedom* at each node. Thus, since displacement at *each* node results in one set of equations, like equations [1, 2, 3] above, it is logical to conclude that two sets of equations result for the element shown in Fig. 3. (one set of equations for each node). If one were to extrapolate the above observation to all seven elements, each with two nodes, for the model shown in Fig. 2, it is evident that:

(7 elements)*(2 nodes/element)*(1 degree of freedom/node) = 14 degrees of freedom [4]

A mathematical solution for the 14 degrees of freedom (displacements) requires simultaneous solution of fourteen equations to solve this very simple problem. The following section examines other element types currently available in COSMOSWorks.

COSMOSWorks Elements

Having established a fundamental understanding of nodes, elements and degrees of freedom, we next introduce the two types of elements available within COSMOSWorks. One is a *Solid* element, the other is a *Shell* element. Each of these element types is available as either a first-order or a second-order element. Descriptions of each element type follow.

Solid Elements

The majority of components analyzed by finite element methods are 3-dimensional models based on solid geometry used to define boundaries of a part or assembly. In this context, *solid* refers to parts or assemblies that have significant volume or thickness relative to other component dimensions. The COSMOSWorks *solid* elements used to model this type of geometry are named tetrahedral elements. A first-order tetrahedral element, shown in Fig. 4, is comprised of six straight sides, four flat faces and four nodes that then join edges at each of its four corners. All sides of first-order elements remain straight and flat after deformation. First-order elements are also called "draft quality" elements.

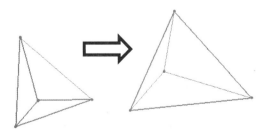

Figure 4 – First-order tetrahedral element before and after deformation.

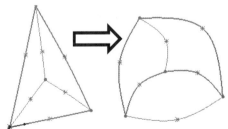

Figure 5 – Second-order tetrahedral element before and after deformation.

I-3

A second-order tetrahedral element, illustrated in Fig. 5, is characterized by the addition of another node on each edge of the element for a total of ten nodes per element. These additional nodes, sometimes called "mid-side nodes," permit edges and surfaces to conform and deform in a second-order (curved) manner. The obvious advantage of second-order elements is their ability to provide better mapping of curved surfaces. And a better fit of elements ensures improved modeling of deformations, strains, and hence stresses computed within modeled parts. Whether a first-order (draft quality mesh) or second-order (high quality mesh) is used, the overall mesh density (i.e., total number of elements) is essentially the same for identical model geometry. However, second-order elements yield better results at the expense of greater computational resources.

Solid Element Degrees of Freedom

Both first-order and second-order tetrahedral elements within COSMOSWorks allow three degrees of freedom at each node. These degrees of freedom permit *displacements* in the X, Y and Z directions. The number of degrees of freedom for each element type is summarized below.

First-order element –
 (3 degrees of freedom/node) * (4 nodes/element) = 12 degrees of freedom/element   [5]

Second-order element –
 (3 degrees of freedom/node) * (10 nodes/element) = 30 degrees of freedom/element   [6]

Recalling the relationship between degrees of freedom and the corresponding number of simultaneous equations requiring solution for each degree of freedom, it is easily seen that computational intensity increases dramatically as model size and order of the element type increases. Although more computationally intensive, second-order elements yield more accurate results and are typically recommended.

Shell Elements

The second type of element available within COSMOSWorks is the *shell* element. As implied by its name, *shell* elements are primarily used to analyze thin-walled components such as sheet metal parts or other thin parts (surfaces) regardless of material. Examples of components that might be modeled using shell elements include a motorcycle gas tank or fenders, beverage containers, thin-wall pressure vessels, many plastic parts or the oil pan on your car. Although the above examples may appear to present clear-cut differences between uses for *solid* and *shell* elements, in reality there exist numerous situations where a "thin" member can be modeled equally well using either type of element. Usually the nature of desired results dictates the choice of element type.

As with solid elements, *shell* elements are also available as first-order "draft quality" or second-order "high quality" mesh types. Figure 6 illustrates a first-order shell element. This element appears two dimensional with straight sides, a flat surface, and one node at each of its three corners. Figure 7 illustrates a second-order shell element with its additional mid-side nodes yielding a total of six nodes per element.

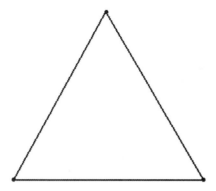 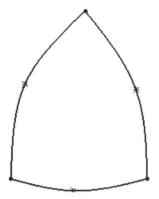

Figure 6 – First-order shell element.

Figure 7 – Second-order shell element showing additional mid-side nodes.

Once again, mid-side nodes permit shell elements to conform to curved geometry of the original component and to better model curvature associated with displacements in deformed parts. Although shell elements are considered two dimensional, a thickness must be associated with them to properly model a part. There are two ways to define a shell mesh. One method creates a shell mesh based on the *mid-surface* of a part while the other approach allows the user to select the surface to be meshed. Since assignment of mesh thickness varies according to how a shell mesh is defined, this topic is investigated further in an illustrative example in Chapter 4.

Shell Element Degrees of Freedom

Because shell elements are two dimensional, it might be presumed that they have fewer degrees of freedom (i.e., displacements) allowed at each node point. However, just the opposite is true because in addition to translational displacements at each node, thin members (shells or membranes) may be subject to bending loads. Therefore, each node of a shell element has six degrees of freedom. They include three displacements (X, Y, Z) and three rotations, one about each of the global X, Y and Z axes.

Meshing a Model

The process of subdividing machine elements into an organized set of nodes and elements is called *meshing,* or creating a *mesh,* on a model of the component to be analyzed. As implied above, mathematical solution of a finite element analysis depends upon sets of simultaneous equations that describe small displacements at the element level. Therefore, subdividing a model into a continuous set of nodes and elements, i.e., meshing a model, is a necessary prerequisite in the solution process. Fortunately, within COSMOSWorks meshing occurs automatically. However, users have the ability to apply mesh controls as is investigated in an example in Chapter 3.

To understand this process it is helpful to outline the progression from an actual part to a meshed finite element model. This process is illustrated in Figs. 8, 9 and 10. Typically,

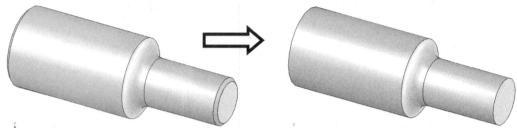

Figure 8 – Solid model of an actual machine shaft to be meshed prior to FEA.

Figure 9 – Solid model of shaft modified to simplify its geometry prior to meshing. (two end chamfers are deleted)

actual part geometry is simplified to reduce or eliminate geometric features that have little or no impact on the ensuing analysis. This step is known as "defeaturing" the model. In this example, 45° chamfers are removed from both ends of the shaft shown in Fig. 8, which leads to the simplified model illustrated in Fig. 9. This may seem like a small change, but an accumulation of similar small changes made to a complex part or assembly can significantly reduce model complexity and hence solution time. Of course, the user must use good engineering judgment when weighing whether or not simplifying changes alter important aspects of the model.

Figure 10 shows the simplified shaft model after meshing it using first-order, solid tetrahedral elements. The model below consists of approximately 1543 nodes and 7192 elements. Automatic meshing of this part was accomplished in less than four seconds using COSMOSWorks 2008 and a Pentium 4® PC.

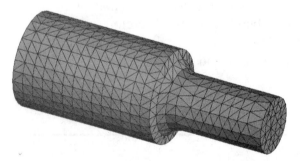
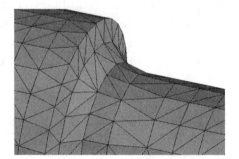

Figure 10 – Meshed model of the shaft using first-order (draft quality) tetrahedral elements.

Figure 11 – Close-up of the fillet reveals the approximate nature of first-order tetrahedral elements.

Figure 11 shows a close-up view of the fillet at the transition between small and large shaft diameters. This image reveals the approximate nature of first-order tetrahedral element edges (straight edges) when used to model a curved surface. The modeling errors that are visible when first-order elements are used contribute to larger approximations in computed results of a finite element analysis. For this reason, second-order (high quality) elements are recommended to achieve more accurate results.

Remaining steps of a finite element analysis are analogous to steps involved when solving stress related problems by long-hand methods. In particular, loads and restraints are applied to the model (like a free-body diagram) and material properties are selected so that stress levels can be compared to material strength at the conclusion of a solution. Because these aspects of problem solution are more familiar to individuals who have completed a fundamental Mechanics of Materials course, discussion of the equivalent finite element steps is deferred to actual examples. For these reasons, discussion below examines and defines various aspects of the COSMOSWorks user interface.

Introduction to the COSMOSWorks User Interface

An overview of the COSMOSWorks user interface is provided below. Perhaps the most awkward part of getting started using SolidWorks, COSMOSWorks or any other complex software program is finding your way around in a new software work environment. Because some users might be new to both COSMOSWorks *and* SolidWorks, this introduction begins by providing orientation to the SolidWorks work environment, also known as a Graphical User Interface or GUI.

Proficient SolidWorks users can skip the SolidWorks orientation section below. However, the toolbars shown in Figs. 12 and 15 should be examined. These toolbars are important because (a) many SolidWorks icons are also useful when working in COSMOSWorks, and (b) geometry of most COSMOSWorks examples in this text are fairly simple and users will maintain or enhance their SolidWorks skills by building each model from scratch.

Orientation and Set-up of the SolidWorks Work Environment

First time users of the combined SolidWorks/COSMOSWorks software must become acquainted with both the vocabulary and the location of items in the work environment. Thus, we begin by investigating how to add or delete menus to the default SolidWorks graphical user interface. The default SolidWorks screen is shown in Fig. 12. Most menu items are inactive (grayed out) at this time.

Figure 12 – SolidWorks start-up screen showing default menus at top and sides of screen.

Depending upon your work environment, particularly at public access computers, the default work environment may have been altered by previous users. Therefore, this

Analysis of Machine Elements using COSMOSWorks

section outlines how to customize the screen to suit your personal preferences. However, before adding or deleting toolbars, it is appropriate to open a new part on the screen even though that part will not be used. To open a new part, proceed as follows.

1. Begin by moving the cursor over the  button located at the top-left of the screen. This action opens the main menu illustrated in Fig. 13.

Figure 13 – The SolidWorks button provides access to the **File**, **View**, **Tools** and **Help** items in the main menu.

2. For those who appreciate the convenience of always having these main menu selections remain visible, a "push-pin", circled in Fig. 13, is provided at the right-end of menu. Click this icon to continuously display the main menu.

3. From the SolidWorks main menu, click **File**, and from the pull-down menu select **New…**. The **New SolidWorks Document** window opens. Within this window, click the **Part** icon. It is highlighted (appears a darker shade) by default.

4. Click **[OK]** to accept this *new* part. The **New SolidWorks Document** window closes and the default SolidWorks screen should appear as shown in Fig. 14.

Various areas of the SolidWorks work environment are identified by labels appended to Fig. 14. Examine these labels to become familiar with nomenclature used to identify portions of the graphical user interface.

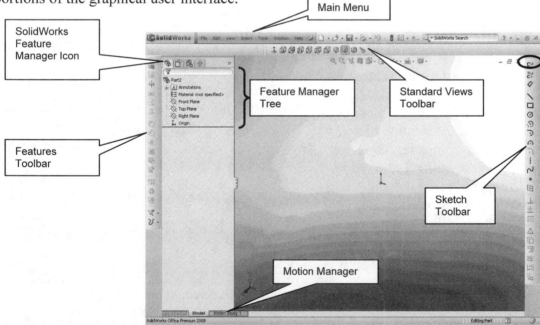

Figure 14 – Definitions applied to basic portions of the SolidWorks screen image. Several default toolbars are shown.

The next task leads you through the addition of various toolbars to the screen and the modification of some existing toolbars.

5. From the main menu, click **View**. Then from the pull-down menu select **Toolbars ▶**. This action opens the Command Manager pull-down menu shown in the partial screen image of Fig. 15. *CAUTION: Two **Toolbars** options exist!*

The **Standard View** icon, highlighted in Fig. 15, corresponds to the **Standard Views** toolbar that already appears at the top of the graphics window in Fig. 14. Note that the **Standard Views** icon appears depressed (darker shading) thereby indicating it is active. *NOTE: It may be necessary to scroll down the list to locate the **Standard Views** icon.*

Although the following steps detail how to add toolbars to the menu, be aware that most necessary toolbars already appear on the screen due to default settings within SolidWorks. Thus, to gain practice with this procedure, follow steps below to *turn off* the toolbars and subsequently to *turn them back on*.

6. Click **Standard Views** (see arrow adjacent to menu in Fig. 15). The menu closes and the **Standard Views** toolbar is deleted from the top of the screen.

7. To open the Command Manager pull-down menu again, use the following shortcut. Right-click anywhere in the light gray area on the top menu bar. This shortcut opens the Command Manager menu.

8. Click to select **Standard View**. The Command Manager menu closes and the **Standard View** toolbar is again added to the top of the screen, shown in Fig. 16.

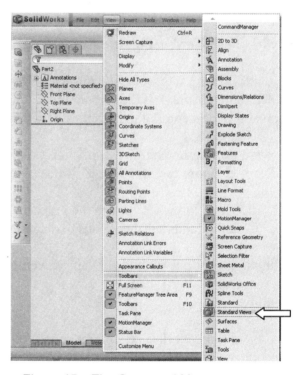

Figure 15 – The Command Manager menu from which toolbars are selected for display.

9. To gain familiarity with this process, repeat steps 5 through 8, but this time select the **Sketch** toolbar. This toolbar is typically placed along the right edge of the graphics window. If it does not appear there initially, its location can be adjusted by clicking-and-dragging a toolbar drag handle. A typical drag-handle is circled at the top of the **Sketch** toolbar in Fig. 14.

Finally, observe the icons appearing at the top-middle of the graphics screen as shown in Fig. 16. These default icons are several of the more frequently used icons within SolidWorks. However, since this text focuses on using COSMOSWorks, we next revise this list of icons.

Figure 16 – Default icons appearing at top-middle of the start-up screen

10. Begin by right-clicking the **Section View** icon circled in Fig. 16. A pop-up menu appears.

11. Within the pop-up menu, click to remove the check mark "✓" adjacent to the ☐ **Section View** icon.

12. Because the menu remains open, also click to *un*-check ☐ **Display Style**, ☐ **Hide/Show Items**, ☐ **Apply scene**, and ☐ **View Settings**. These icons are underlined in Fig. 16.

In the following step, several icons useful for manipulating a model on the screen are added to the display. If the pop-up menu was allowed to close, simply right-click any icon at the top-center of the graphics viewport to open it again.

13. Within the pop-up menu, click to select the following icons: ☑ **Zoom In/Out**, ☑ **Zoom to Selection**, ☑ **Rotate view**, ☑ **Roll view**, and ☑ **Pan**.

14. Click anywhere within the graphics screen to close the pop-up menu. This action adds selected icons to the default display.

Although short-cut methods of using several of the view icons were introduced in the Preface (see **Table 1**, page ix), new users may find the visual display of icon capability helpful.

Seasoned SolidWorks users might choose to place additional toolbars on the screen, or customize those now appearing by adding or removing specific icons. The procedure to do this is much the same as adding or removing icons in any Windows® program. For this reason it is not uncommon for slight differences of screen images to occur. Some blank areas on the upper toolbar menu will be filled in with COSMSOWorks toolbar icons as demonstrated in the next section. Without making further changes to the work environment, proceed directly to the next section.

Orientation to the COSMOSWorks Work Environment

Following procedures equivalent to those outlined above, essential COSMOSWorks toolbars are added to the screen next. However, it is first necessary to deal with the possibility that the COSMOSWorks component of SolidWorks has not yet been added to the work environment. Thus, proceed as follows to activate COSMOSWorks.

1. In the main menu, click **Tools** and from the pull-down menu select **Add-Ins…**. The **Add-Ins** window opens as shown in Fig. 17.

2. Within the **Add-Ins** window, click to place a check mark to the *left* of ☑ **COSMOSWorks 2008** and in the **Start Up** column located to the *right* of **COSMOSWorks 2008**.

3. Click **[OK]** to close the **Add Ins** window.

This action adds **COSMOSWorks** to the main menu and also adds the COSMSOWorks Manager icon to the top of the manager tree located to the left of the graphics screen as shown on the partial screen image in Fig. 18.

4. Click the **COSMOSWorks** icon to change appearance of the COSMOSWorks Manager as shown in Fig. 18.

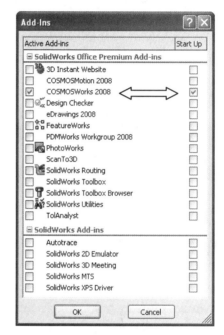

Figure 17 – Activating COSMOS-Works within the **Add-Ins** window.

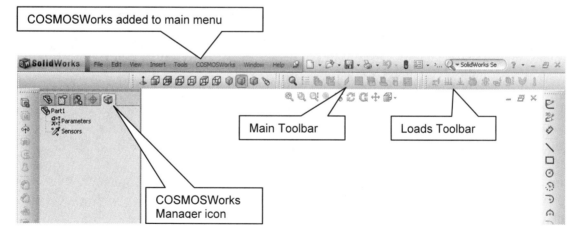

Figure 18 – Work environment modified by the addition of COSMOSWorks and two COSMOSWorks toolbars.

Next, two COSMOSWorks toolbars are added as illustrated in Fig. 18.

I-11

5. Right-click anywhere in the gray area of the upper menu. The **Command Manager** menu opens. A partial view of this menu is shown in Fig. 19.

6. Click to select **COSMSOWorks Loads** and the loads toolbar is added above the graphics area in Fig. 18.

7. Repeat step 5 to open the **Command Manager** menu. This time, select **COSMOSWorks Main** and the main toolbar is also added to Fig. 18.

8. Click-and-drag the toolbar drag-handle (located at left of each toolbar) to move these new toolbars to convenient locations above the graphics screen.

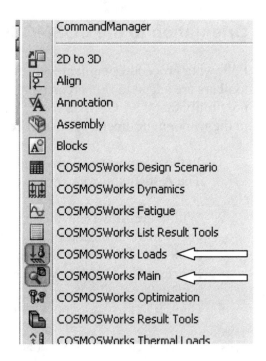

Figure 19 – Command Manager lists toolbars available in COSMOSWorks.

Property Managers and Dialogue Boxes

Property managers and dialogue boxes are, perhaps, the most frequently referred to items when developing a finite element analysis within COSMOSWorks. The importance of these two interfaces cannot be overstated because the specifics of each part of the finite element modeling process are defined within them. However, these interfaces are only encountered when defining an actual analysis. For this reason, this section only attempts to define their location on the screen and provide basic insight to their general role in an analysis. Because an actual analysis is not currently being performed, this section is descriptive in nature only and does not require user interaction.

When opened, a *property manager* is typically located at the left side of the graphics screen, shown boxed in Fig. 20. Numerous property managers exist and each is somewhat unique. However, the general observations included below apply to each. The property manager, shown boxed in Fig. 20, is enlarged in Fig. 21.

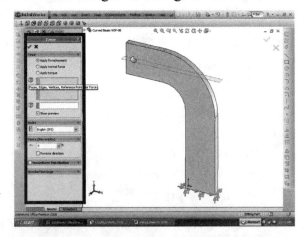

Figure 20 – Image showing property manager location in the COSMOSWorks graphical user interface.

Figure 21 shows an example of the **Force** property manager with its name prominently displayed at the top. This property manager is used when defining different types of loads applied to a finite element model.

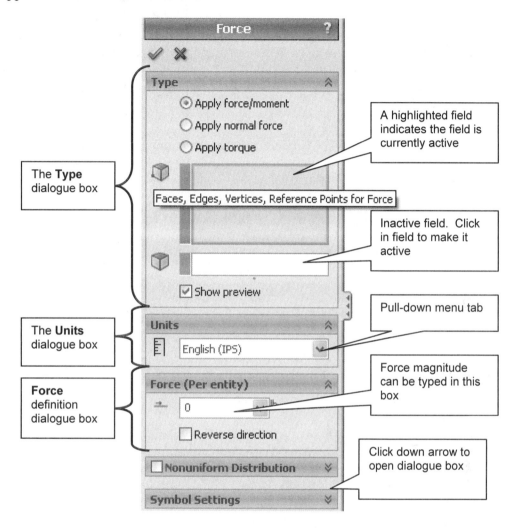

Figure 21 – Overview of features common to property managers and dialogue boxes.

Each property manager contains several *dialogue boxes*. Five different dialogue boxes appear in the **Force** property manager; they are: **Type**, **Units**, **Force (Per entity)**, **Nonuniform Distribution** and **Symbol settings**. Dialogue boxes group related items in a single location and their names help the user quickly locate needed items.

For example, the **Type** dialogue box lists various types of forces, moments, or torques that can be selected by clicking within the ⊙ symbol to the left of a specific load type. Also within the **Type** dialogue box are two *fields*. Moving the cursor over a field or the icon adjacent to a field, causes specific field information to appear much like occurs for other Windows® icons. In Fig. 21, the name of the upper field is displayed as "**Faces, Edges, Vertices, Reference Points for Force**". This field is also highlighted (here shown in gray), which indicates it is currently "active" and awaiting user input.

Many dialogue boxes contain pull-down menus, which can be accessed by clicking the symbol adjacent to the box. For example, the **Units** dialogue box in Fig. 21 permits the user to switch between **English (IPS)**, **SI** or **Metric (G)** units. Some boxes, such as the **Force (Per entity)** dialogue box, permit the user to type a force value into a data box. And finally, as is often the case, insufficient space is available to display the contents of all dialogue boxes simultaneously. Therefore, only the titles of some dialogue boxes appear, but the user must click the symbol to display their contents. Such is the case for the **Nonuniform Distribution** and the **Symbol Settings** dialogue boxes at the bottom of Fig. 21.

This completes the introduction to underlying fundamentals of finite element analysis and the SolidWorks/COSMOSWorks graphical user interface. Familiarize yourself with this work environment so that you are able to work comfortably within it.

Examples focusing on the application of COSMOSWorks to solve finite element modeling problems begin in the next chapter.

CHAPTER #1

BASIC STRESS ANALYSIS USING COSMOSWorks

This example is intended to familiarize first-time users of COSMOSWorks with basic software capabilities. Particular aspects of the software illustrated in this example are listed below. This example also serves as a *model* for subsequent Finite Element Analysis (FEA) problems because, once mastered, the sequence of solution steps is fairly consistent. These steps closely parallel the first six learning objectives outlined below.

Learning Objectives
Upon completion of this unit, users should be able to:
- Create and execute a linear, static Finite Element Analysis (FEA) using COSMSOSWorks. Within COSMOSWorks this process is named a *"Study."*

- Assign *Material Properties* using the COSMOSWorks material editor.

- Apply *Restraints* and *Loads* to a model.

- Use the default *Mesh* definition to subdivide a part into nodes and elements.

- Execute a standard *Solution* to a FEA problem.

- Selectively view appropriate stress *Results* of a Finite Element Analysis.

- Use the *Probe* feature to create graphs of stress variation within a part.

- Develop insight into practical decisions that influence how a *Study* is defined to obtain desired results.

Problem Statement
This example presumes it is desired to determine stresses in the reciprocating cam follower illustrated in Fig. 1. In particular, stresses to be determined are those in the circled region near the upper end of the cam follower where it passes through its support in the frame.

Positions of the cam and follower shown in Fig. 1 and Fig. 2 are assumed to correspond to the locations of maximum cam pressure angle and maximum dynamic load on the cam follower. Although not shown here, a dynamic analysis should be performed using COSMOSMotion and the resulting dynamic loads used here.

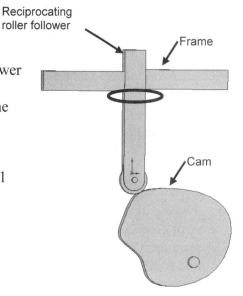

Figure 1 – Concept sketch showing the region of interest on the cam follower.

Figure 2 shows specific dimensions and loads applied to the cam follower. Force components $F_x = -368$ lb and $F_y = 1010$ lb exerted by the cam on the follower in the X and Y directions are applied on the roller-pin at the bottom of the follower. Because a static stress analysis is to be performed, the upper-end of the cam follower is considered "fixed," analogous to the fixed end of a cantilever beam, and corresponding reaction forces R_x, R_y, and a resisting moment M_z are shown at the upper support. We will soon discover these end conditions are not possible if solid tetrahedral elements are used to model the cam follower.

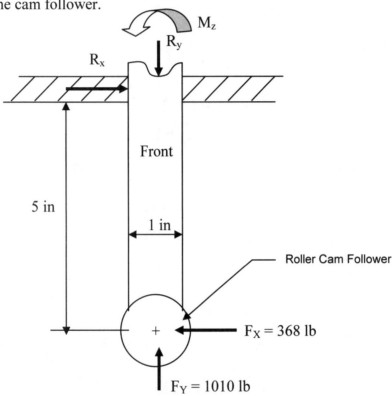

Figure 2 – Two-dimensional model of the reciprocating cam follower. The *Front* surface is labeled for later reference. The follower stem is ½ in. deep into the page.

Design Insight

Although finite element programs are powerful computational tools, you should be aware that engineering decisions are necessary to advance the design process. In this example, at least three significant engineering assumptions are applied. They are:

- The list of possible materials must be narrowed to a group that provides a strong yet heat-treatable class of steels. Heat treatment is important to produce a hard, long wearing surface for contact between the cam follower and frame.

- Because the focus of this analysis is on stresses at the *upper* end of the cam follower, modeling of the roller and its pin connection to the follower can be neglected at a considerable saving of effort.

Basic Stress Analysis

> - Assuming the upper-end of the cam follower is "fixed" results in significant simplification for this initial example. (i.e. a static rather than a dynamic analysis results).

Creating a Static Stress Analysis Study

> Before proceeding, download and unzip chapter examples and end-of-chapter problems from the publisher's web site:
> **http://www.schroff1.com**

1. Open SolidWorks by making the following selections. These steps are similar to opening any Windows program. (*Note:* The "/" is used to separate successive menu selections.)

   Start / All Programs / SolidWorks 2008

2. When SolidWorks opens, select **File / Open**. Then use procedures common to your computer environment to open the file named **Cam Follower (Part)**.

If a pop-up SolidWorks window appears and states: **The following documents will be converted when saved: Cam Follower (Part).SLDPRT**, click **[OK]** to close the window.

NOTE: If **COSMOSWorks** is not listed in the main menu of the SolidWorks screen, then in the main menu click **Tools / Add-Ins...** Next, in the **Add-Ins** window "check" ☑ **COSMOSWorks 2008** and also place a "check" ☑ in the **Start Up** column and click **[OK]** to close the **Add Ins** window. Then proceed as follows.

The **COSMOSWorks License Agreement** appears *(for first-time users only)*. Click **[Accept]**. At this point COSMOSWorks is added to the main menu and a COSMOSWorks icon (a small cube with a "C" and a "W" on its sides) appears at the top of the feature manager located at the left side of the screen. Generically, this window is referred to as the manager tree.

3. Click the COSMOSWorks icon to open the COSMOSWorks Manager shown in Fig. 3

4. In the COSMOSWorks Manager, right-click **Cam Follower (Part)** and from the pull-down menu, select **Study...** The **Study** property manager appears as seen in Fig. 4.

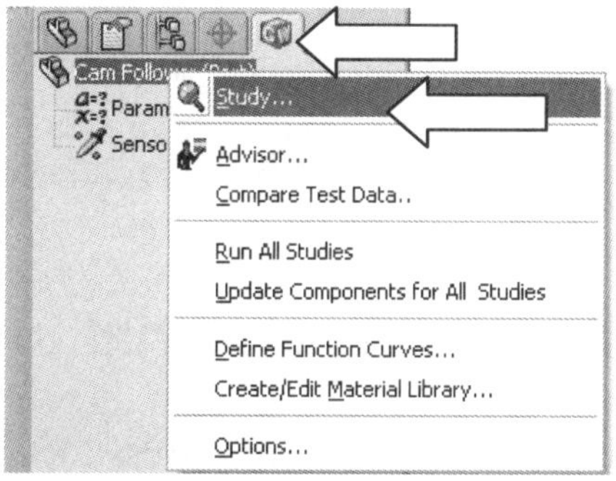

Figure 3 – Opening a **Study...** in COSMOSWorks.

1-3

5. In the upper field of the **Name** dialogue box, replace the name "**Study 1**" by typing a descriptive name. For this example type, "**Cam Follower #1**".

6. Move the cursor over the second field from the top of the **Name** dialogue box. The field identifier appears as **Mesh type** to indicate this field is used to specify the type of mesh to be used. From the pull-down menu select **Solid mesh** (if not already selected). A solid mesh is chosen because the cam follower is a basic three-dimensional part.

7. Finally, in the **Type** dialogue box, verify that the system default **Static** analysis icon appears shaded to indicate it is selected.

Additional analysis capabilities listed below the bold line in Fig. 4 are available in COSMOSWorks Professional. This users' guide focuses on **Static** analysis capabilities found in COSMOSWorks Designer and listed above the bold line.

8. Click **[OK]** ✓ (the green check mark) to close the **Study** Property manager.

Figure 4 – Initial selections shown in the **Study** property manager.

Figure 5 – The COSMOSWorks analysis manager showing various components used to define a Study.

After completing the above steps, an outline of the current Study is created in the COSMOSWorks manager as shown in Fig. 5. The Study name **Cam Follower #1 (-Default-)** appears in bold font. Beneath this name is the **Solids** folder that contains a solid model of the **Cam Follower (Part)**, click the "+" sign to reveal the **Cam Follower (Part)**. Also, icons for the applied **Load/ Restraint**, and the model **Mesh** appear. Additional icons for the **Design Scenario, Contact/Gaps (-Global: Bonded-)**, and **Report** are listed, but they are not considered in this example.

Assigning Material to the Model

Because the outline for this study, seen in Fig. 5, lists steps for developing a finite element model in a logical order, this sequence is followed from top to bottom in this and subsequent examples. However, it is worth noting that steps shown in the outline can be

Basic Stress Analysis

executed in any order. Begin by defining the material of which the cam follower is made. To do this, proceed as follows.

1. In the COSMOSWorks manager right-click the **Solids** folder; see Fig. 6, and from the pull-down menu, select **Apply Material to All…**. The **Material** window opens as shown in Fig. 7.

2. Near the top-left of the **Materials** window, beneath **Select material source**, do the following:

    a. Click to select ⊙ **From library files**.

    b. From the pull-down menu beneath **From library files**, select the **cosmos materials** option (if not already selected).

Figure 6 – Selecting the part to which material properties are assigned.

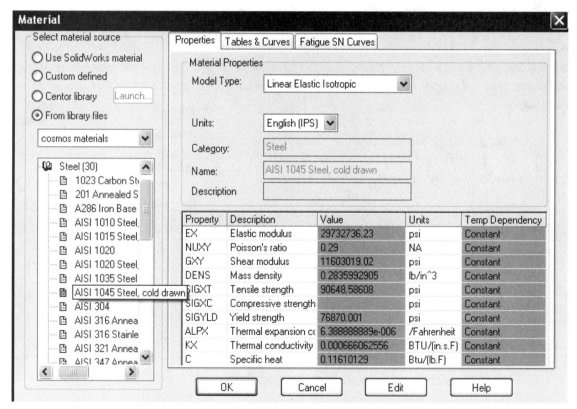

Figure 7 – Material properties of a part are specified in the **Material** window.

   c. Click the plus sign next to "+" **Steel (30)** and scroll down the list to select **AISI 1045 Steel, cold drawn**. Immediately, properties of 1045 steel appear in the right half of the table.

1-5

Analysis of Machine Elements using COSMOSWorks

    d.   If material properties are listed in SI units, click the **Units:** pull-down menu. From the list, select the **English (IPS)** system of units, where IPS indicates units of inches, pounds, and seconds.

Familiarize yourself with information in the table by reading property names listed in the **Description** column and corresponding magnitudes in the **Value** column. For example, **Yield Strength = 76870.001 psi** for AISI 1045 cold drawn steel. Obviously, greater precision is indicated than is truly known for this material. This is because S.I. units are default within COSMOSWorks and extra digits frequently occur when values are converted to English units. Examine other values listed in the table to become familiar with data available in the material library.

    3.   Click **[OK]** to close the **Material** window. In the COSMOSWorks Manager tree, notice that a check mark "✓" now appears on the **Cam Follower (Part)** and the **Solids** folder. This mark indicates that material properties have been defined.

Applying Restraints

Adequate restraints must be applied to stabilize and support the model as it is supported in its actual application. In this example, the top surface of the model, i.e., the location where the cam follower enters the frame, is assumed to be "fixed." This restraint is consistent with a cam follower whose upper end is located in a near-zero clearance slot in the frame and for which a static analysis is assumed. Define restraints for the model as follows.

    1.   In the COSMOSWorks manager tree, right-click **Load/Restraint**, then from the pop-up menu select **Restraints…**.

        The **Restraint** property manager appears as shown in Fig. 8.

    2.   Under restraint **Type**, select **Immovable (No translation)** from the pull-down menu. This option resists translations in X, Y and Z directions. Recall that tetrahedral elements used to model solids allow only three degrees of freedom (three translations) at each node. Thus a "Fixed" restraint *cannot* be applied to this model.

Figure 8 – Selections made in the **Restraint** property manager.

As with other Windows operations, placing the cursor over an icon causes a brief description of the icon to be displayed. Placing the cursor on the icon located to the left of the light blue-colored field in the **Type** dialogue box reveals that **Faces, Edges,**

Basic Stress Analysis

Vertices for Restraint can be selected as entities to which restraints can be applied. In this example the top surface of the cam follower is selected as described next.

The light blue color indicates that the **Faces, Edges, Vertices for Restraint** field is *active* and waiting for the user to specify what part of the model is to be designated as **Immovable**. If selecting the top surface of the model is difficult due its orientation or size, use various graphics controls to rotate, and/or zoom-in on the top surface. If rotate or zoom icons are used, press **[Esc]** to return to the standard cursor (pointer).

3. Move the cursor over the model and when the top *surface* is indicated by a shaded square next to the cursor, click to select the surface. Because the ☑ **Show preview** box is checked (by default) the top surface will appear highlighted and **Immovable** restraint symbols appear as shown in Fig. 9. Also, the notation **Face<1>** now appears in the **Restraint** property manager shown in Fig. 8.

Figure 9 – **Immovable** restraint applied to top surface of the cam follower model.

If an edge, vertex or other surface of the model is selected incorrectly, right-click the incorrect item in the dialogue box field and select **Delete**. Then repeat step #3.

4. To change color or size of the restraint symbols, click the down-arrow ⌄ to open the **Symbol Settings** dialogue box, shown in Fig. 8. Click the **[Edit Color...]** button to open a color palette. Select the desired color and click **[OK]**. In this example you are encouraged to leave restraint color as pre-defined because additional symbols applied to the model later in this analysis use different default colors to differentiate restraints from applied loads. To change symbol size, click the up or down arrows adjacent to the **Symbol Size** spin box. Restraint symbols in Fig. 9 were increased in size to 150 and their color changed for emphasis.

5. Click **[OK]** ✓ (green check-mark) at top of the **Restraint** property manager to accept these restraints and close the window. The above steps render the top face of the cam follower **Immovable** and create an icon named **Restraint-1** beneath the **Load/Restraint** folder in the COSMOSWorks manager.

Aside #1:
Most readers might question why **Immovable** restraints, rather than **Fixed** restraints, are specified for the *fixed* end-condition on top of the cam follower. To understand this action, recall the "Introduction" to this user guide where various COSMOSWorks element types are described. Of those elements, the three-dimensional tetrahedron is characterized by corner node points whose motion is restricted to translations in the X, Y and Z directions giving it only three degrees of freedom at each node. Therefore, because the **Immovable** restraint imposes zero displacements in these three directions, it effectively eliminates all possible motion of restrained nodes for this type of element.

Analysis of Machine Elements using COSMOSWorks

> The **Fixed** restraint, on the other hand, not only prohibits node translations in the X, Y and Z directions, it also restricts (prevents) rotational displacement about the X, Y and Z axes at each restrained node. Since solid tetrahedral elements do not have a rotational degree of freedom, it makes no sense to apply the **Fixed** restraint. **Fixed** restraints are typically applied to shell elements that allow three-dimensional translation and rotation
> at each node. A shell mesh will be investigated in Chapter 4.
>
> **Aside #2:**
> Application of **Fixed** restraints to the current problem would also yield valid results. However, it is instructional to recognize differences between restraint types, hence the above discussion. If a **Fixed** restraint were applied to this example, the rotational restraints would simply be ignored since they do not apply to tetrahedral elements.

Applying Loads

Following the sequence of steps suggested in the COSMOSWorks manager, we next apply loads (i.e., externally applied forces) to the model. Proceed by defining the X and Y force components one at a time that act at the bottom center of the cam follower. Begin by applying the vertical component of force $F_y = 1010$ lb that acts upward on the bottom of the cam follower. Proceed as follows.

1. In the COSMOSWorks Manager, right-click the **Load/Restraint** folder and from the pull-down menu, select **Force…**.

    The **Force** property manager appears in Fig. 10.

2. Under **Type**, click to select ⦿ **Apply force/ moment**. The upper light-blue colored field is highlighted thereby prompting the user to select the entity to which the force is to be applied. Move the cursor onto this field and observe the prompt that reads, **Faces, Edges, Vertices, Reference Points for Force**.

3. In the graphics area, rotate the model and zoom-in on the bottom face. Notice the line that appears at the center of the bottom surface. As you move the cursor onto this line, the message *(Split Line1)* appears. Click to select this line since it is desired

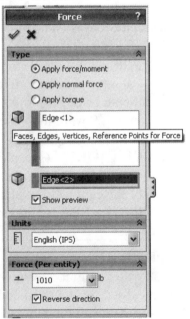

Figure 10–Definition of force F_y normal to bottom surface.

Basic Stress Analysis

to place both X and Y force components at the center of the bottom face. After selecting *(Split Line1)*, **Edge<1>** appears in the top field of the **Type** dialogue box in Fig. 10 and an information "flag" appears adjacent to the model.

4. Next, click inside the bottom field in the **Type** dialogue box to highlight this field. Placing the cursor on this field prompts the user to select a **Face, Edge, Plane, Axis for Direction**. In other words, the next step is to indicate the direction of the Y-force component by selecting a face, edge, or axis in the desired direction. Select any *vertical* **Edge** of the model as shown in Fig. 11. Provided ☑ **Show preview** is checked, force vectors appear at both ends of *Split Line1* and are oriented in the Y direction. Ignore incorrect directions at this time. Also, **Edge<2>** appears in the active field.

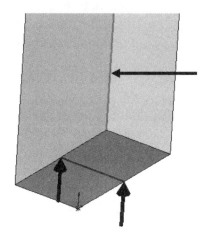

Figure 11 – Vertical force components and the edge used to specify direction of the force acting on Split Line1.

5. Also within the **Force** property manager, use the pull-down menu in the **Units** dialogue box to set **Units** to **English (IPS)** (if not already selected).

6. Finally, in the field beneath **Force (Per entity)**, type **1010**, which is the magnitude of the Y-force component. Examine the direction of vectors appearing in the graphic image and if they are not directed toward the top of the model, check ☑ **Reverse direction.**

7. Click **[OK]** ✓ (green check-mark) to accept this force definition and close the **Force** property manager. COSMOSWorks applies the 1010 lb force to the model and creates an icon named **Force-1** beneath the **Load/Restraint** folder in the COSMOSWorks manager tree.

Next, apply the X-force component ($F_x = -368$ lb) to *Split Line 1* on the bottom of the cam follower. Try this on your own. However, a brief listing of commands is provided in steps 8 through 14 below for those desiring guidance.

8. In the COSMOSWorks manager tree, right-click the **Load/Restraint** icon and select **Force…** The **Force** property manager appears.

9. Under **Type**, select ⊙ **Apply force/moment**.

10. The active (light blue) field **Faces, Edges, Vertices, Reference Points for Force** prompts the user indicate where the force is to be applied. Again select *Split Line 1*. The line is highlighted and **Edge <1>** appears in the **Type** dialogue box.

1-9

11. Click to activate (light blue color) the **Face, Edge, Plane, Axis for Direction** field and proceed to select an edge parallel to the X-direction (i.e. parallel to the 1-inch dimension of the cam follower). Any edge, even a top edge, can be selected as long as the edge is in the X-direction. After selecting an edge, **Edge<2>** appears in the direction field and force vectors in the X-direction appear on both ends of *Split Line1*. Ignore incorrect directions at this time.

12. Set the **Units** field to **English (IPS)**.

13. In the **Force (Per entity)** field, type **368** to define magnitude of the X-force component. If necessary, check ☑ **Reverse direction** to orient force components in the negative X-direction.

14. If the X force component appears as shown Fig. 12, click **[OK]** ✓ to accept this force and close the **Force** property manager.

15. An icon named **Force-2** appears beneath the **Load/Restraint** folder.

Figure 12 – Front view of cam follower showing force components F_x and F_y applied to the split line.

The model is now complete as far as material property, restraint, and force definitions are concerned. The next step is to mesh the model as described in the following section.

Aside #1:
Split lines and their use are frequently encountered in the study of SolidWorks therefore this example does not review their use. However, they are also extremely useful in Finite Element Analysis, and for that reason, they are reviewed in future examples.

Aside #2:
As noted in the **Design Insight** section at the beginning of this example, stresses at the *upper end* of the cam follower are to be investigated. For this reason, the very simplified assumption of force loading at the center of the lower end of the model, where the roller is attached to the follower, might be deemed acceptable. After all, why devote considerable time and effort to model contact stress between the roller-pin and cam follower *if* the focus of analysis is to determine stresses elsewhere in the model? On the other hand, if the focus of this analysis were on stresses in the vicinity of the pin that joins to roller to the cam follower, then details of that geometry must be included in the model. Contact stress between a pin and a hole will be investigated in Chapter 6.

Basic Stress Analysis

Meshing the Model

1. In the COSMOSWorks manager tree, right-click the **Mesh** icon and from the pull-down menu, select **Create Mesh....** The **Mesh** property manager opens as shown in Fig. 13.

2. The **Mesh Parameters** dialogue box shows information about the current mesh (i.e., units, mesh size, and tolerance) in fields near the top of this dialogue box. Global mesh size is indicated in the middle field as **0.135767** in., while mesh tolerance, which expresses the allowable deviation from the specified mesh size, is listed in the lower field. These values are automatically calculated by the software based upon model geometry. As such, these values usually provide sufficient definition of element size to yield acceptable results for an *initial* finite element analysis.

3. Next click the down arrow ⍟ to open the **Options** dialogue box, which appears in the lower portion of Fig. 13. Within this dialogue box all or most of the following selections should already appear as default settings. These settings should produce a good quality mesh. However, verify that the settings are as listed in Fig. 13 and only change them if they differ.

4. Under mesh **Quality:** select ⊙ **High**.

5. Under **Mesher:** select ⊙ **Standard**.

6. Finally, under **Mesher options:** check ☑ **Jacobian check for solid** and verify that **4 point** appears in the accompanying pull-down menu.

Figure 13 – Mesh parameters, such as global size, tolerance and mesh quality can be set or altered in the Mesh property manager.

Although most of the above are system default settings, it is always wise to verify them, particularly on "public access" computers.

7. Click **[OK]** ✓ to accept the default values and close the **Mesh** property manager.

1-11

Analysis of Machine Elements using COSMOSWorks

Meshing starts automatically and the **Mesh Progress** window appears briefly. After meshing is completed, COSMOSWorks displays the meshed model as shown in Fig. 14. Also, a check mark "✓" appears on the **Mesh** icon in the COSMOSWorks manager tree to indicate meshing is complete.

8. To display mesh information, right-click **Mesh** and select **Details…** The **Mesh Details** window opens and is also shown in Fig. 14.

Within this window COSMOSWorks displays information about the current model such as its **Study name**, **Mesh type (Solid Mesh)**, **Mesher Used (Standard)**, …, **Element size**, **Tolerance** values and other data is repeated. Scroll down in this window and notice that approximately 10482 **Total nodes** and 6511 **Total elements** are created for this model. Because the automatic meshing software attempts to create an optimal mesh for each model, the number of nodes and elements may vary slightly between alternate meshings of the same model. Click ☒ to close this window when finished examining its contents.

9. To hide the mesh, right-click **Mesh** and from the pull-down menu select **Hide Mesh**. COSMOSWorks hides the mesh. Conversely, selecting **Show Mesh** in the pull-down menu returns the mesh display on the model. Try this option, but Hide the mesh before continuing.

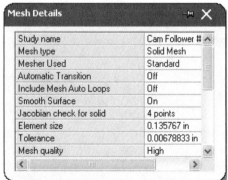

Figure 14 – Cam follower with mesh and boundary conditions illustrated. Also shown is the **Mesh Details** window where mesh information can be reviewed.

Aside:
As noted earlier, it is permissible to define material properties, restraints, loads and create the mesh in *any order*. However, all these *necessary* steps must be completed prior to running the Solution portion of a study.

Basic Stress Analysis

Running the Solution

After the model has been completely defined, we are ready to proceed to the *Solution* process. This is the second major portion of a finite element program. It is where the numerous equations that define the Study are solved. For all of its complexity, this portion of a Finite Element Analysis is, perhaps, most deceiving in terms of its seeming simplicity for software users. Time required for an analysis can vary from several seconds to several hours depending upon overall model complexity and computer hardware used. Most examples in this text should solve in a matter of seconds or a few minutes at most. To solve the current example, proceed as follows.

1. To run the analysis, right-click **Cam Follower#1 (-Default-)** located in the COSMOSWorks manager. Refer to highlighted text in Fig. 15.

2. From the pull-down menu, select **Run** and the solution process begins automatically. A window that tracks progress of the Solution appears, but due to the small size and simplicity of this example, it is displayed only briefly.

After successful solution of a static analysis, COSMOSWorks creates a new folder, named "**Results**," at the bottom of the COSMOSWorks manager tree. This folder should contain the three sub-folders shown in Fig. 15. These three sub-folders contain default plots resulting from a finite element analysis of the current model. If these folders do *not* appear, follow steps (a) through (e) outlined below, otherwise skip to the next page.

a) Right-click the **Results** folder and from the pull-down menu, select **Define Stress Plot...** The **Stress Plot** property manager opens.

b) In the **Display** dialogue box, select **VON: von Mises Stress** from the pull-down menu.

c) Also in the **Display** dialogue box, select **psi** from the **Units** pull-down menu.

d) Click **[OK]** ✓ to close the property manager and immediately a plot of the von Mises stress is displayed in the graphics area.

e) Repeat steps (a) through (d), but in step (a) substitute **Define Displacement Plot...** or **Define Strain Plot...** in place of **Define Stress Plot...**. Also, alter the **Units** field to **in** (inches) for the displacement plot.

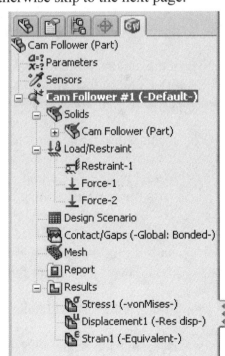

Figure 15 – Folders created as part of the Solution process. Each folder contains solution information indicated by its name.

1-13

Examination of Results

The outcome of this analysis, in the form of graphs, plots and data, can be viewed by accessing results stored in the **Results** folders. These results are the typical goal of a finite element analysis. It is where validity of finite element results should be investigated by cross-checking them against manual calculations or other verifiable experimental or reference results. *Checking computed results is a necessary step in good engineering practice!*

Default COSMOSWorks Graphical Results

1. If a plot of **Stress1 (-vonMises-)** is already displayed because steps (a) through (d) of the previous page were performed, then continue at the next step. Otherwise, in the COSMOSWorks manager, right-click the **Stress1 (-vonMises-)** folder and from the pull-down menu select **Show**. Alternatively, double-click **Stress1 (-vonMises-)** to display the plot.

> **Aside:**
> In this user guide, images of stress within the loaded model are referred to as "stress contour plots" or simply "plots." On the other hand, when an image shows the relationship between two variables in the form of an X-Y line graph, the image is referred to as a "graph."

Either of the actions in step 1 causes a deformed view of the cam follower to appear along with a plot of the von Mises stress distribution, displayed in multiple colors, on the model surface. Von Mises stress is the default stress plot produced by COSMOSWorks.

2. Default information provided near the upper-left of the graphics screen includes:

- **Model name:** ← Contains the name of the SolidWorks model opened from the parts file at the beginning of this example; it is named "**Cam Follower (Part).**"

- **Study name:** ← Name selected and typed into the **Study** property manager at the start of this example. If previous instructions were followed precisely, this name should appear as "**Cam Follower #1**".

- **Plot type:** ← Should indicate "**Static nodal stress Stress1**"

- **Deformation Scale:** ← To illustrate a deformed shape, COSMOSWorks scales the maximum deformation of the model to 10% of the diagonal of a bounding box around the model (this box is not visible). A number representing the magnitude of this scale factor 39.3068 (or similar value) appears on this line. The deformed model is provided for visualization purposes only and does *not* represent the actual magnitude of part deformation.

- Also included is a color-coded stress scale. Stress magnitudes at various locations throughout the model can be determined by matching colors to those of the color-coded stress scale. Red traditionally corresponds to high stress while dark blue corresponds to algebraically lower stress magnitudes. The default location for this scale is on the right of the screen; however, its location can be changed.

Results Predicted by Classical Stress Equations

It is assumed that some users may not yet be familiar with von Mises stress.[1] Therefore, discussion below digresses to examine *other*, more fundamental, stresses that occur within the cam follower model. To accomplish this, a brief but thorough solution to the current problem is included based on classical stress analysis equations.

Begin by recalling the original free-body diagram of the cam follower, Fig. 2. That figure revealed the part is subject to force components in both the X and Y directions on its lower end. The upward acting force component $F_y = 1010$ lb causes an axial compressive stress given by $\sigma_y = F_y/A$ in the Y-direction. Although this stress is shown near the lower end of the cam follower in Fig. 16, classic stress equations assume it is *uniformly distributed* throughout the model from top to bottom. Similarly, the X-component of force, $F_x = -368$ lb, which acts perpendicular to the length of the cam follower, causes bending stress given by $\sigma_y = Mc/I$ shown at the top of the model in Fig. 16. Bending stress is maximum at the top of the model due to maximum length of the moment-arm (5.0 in) at this location. Briefly review stress calculations included on Fig. 16 before proceeding.

Because all stresses depicted on Fig. 16 act in the Y-direction, they can be combined by simply adding magnitudes of axial and bending stresses provided proper $\pm$ signs are included. Combining these stresses yields the following results:

Stress at point A is in compression due to both axial and bending stresses, thus-
  Stress at point A: $(\sigma_y)_A = \sigma_{Axial} + \sigma_{Bending} = (-2020) + (-22080) = -24,100$ psi

Stress at point B is in compression due to axial stress, but bending stress is zero on the neutral axis, thus-
  Stress at point B: $(\sigma_y)_B = \sigma_{Axial} + \sigma_{Bending} = (-2020) + (0) = -2,020$ psi

Stress at point C is compressive due to axial stress, but in tension due to bending stress. Because the "+" bending stress is greater than the "-" axial stress, the resultant stress at point C is-

  Stress at point C: $(\sigma_y)_C = \sigma_{Axial} + \sigma_{Bending} = (-2020) + (+22080) = +20,060$ psi

---

[1] It is presumed that use of this COSMOSWorks user guide will be introduced near the beginning of a Design of Machine Elements or Mechanics of Materials course. However, most traditional textbooks on these subjects delay introduction of vonMises stress until later chapters. For this reason, early examples in this user guide involve stresses that are more familiar to individuals who have completed a fundamental mechanics of materials course.

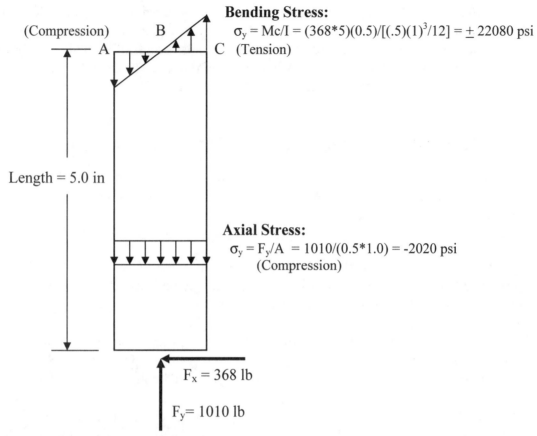

Figure 16 – Schematic of loading and distribution of bending stress (compression on left-side and tension on right-side) and axial stress (compression) in the Y-direction on the cam follower determined using classic stress equations.

COSMOSWorks Results for Stress in Y-Direction

The above results are next compared with those determined using finite element analysis. To do so, it is meaningful to produce a plot of normal stress in the Y-direction (i.e., σ_y). Proceed as follows.

1. In the COSMOSWorks manager tree right-click the **Results** folder and from the pull-down menu select **Define Stress Plot…**. The **Stress Plot** property manager opens as shown in Fig. 17.

2. In the **Display** dialogue box, click to open the pull-down menu adjacent to the **Component** field. Initially this field shows that the **VON: von Mises Stress** is selected for display. From the list of stresses available in the pull-down menu, locate and highlight **SY: Y Normal Stress**. This selection identifies normal stress in the Y-direction, commonly represented by σ_y, as the stress to be displayed in a new plot.

Basic Stress Analysis

For future reference it is informative to observe all the other stresses available for analysis within the **Component** field. Briefly return to the pull-down menu and note the list of thirteen stresses included there. Although names, rather than Greek symbols are listed, observe what stresses it is possible to select from the following *partial* list of stresses available.

$\sigma_x, \sigma_y, \sigma_z$ = Normal stresses in X, Y and Z directions listed as **SX, SY, SZ**.

$\tau_{xy}\ \tau_{xz}\ \tau_{zy}$ = Shear stresses on X, Y, and Z planes, listed as **TXY, TXZ, TYZ**.

$\sigma_1, \sigma_2, \sigma_3 = 1^{st}, 2^{nd}$, and 3rd Principal stresses listed as **P1, P2, P3**.

3. Beneath the stress **Component** field, verify that units are set to **psi**. If not, select **psi** from the pull-down menu.

4. Because it is often convenient to turn off the deformed shape when viewing stresses within a model, click to clear the check mark "✓" from the ☐ **Deformed Shape** dialogue box. Suppressing display of the deformed shape also proves helpful when using the **Probe** feature later in this example.

5. Open the **Property** dialogue box ⌄, and click to place a check mark next to ☑ **Include title text** and type a descriptive title such as **Sigma-Y,** *Your name*. A descriptive title serves both to identify *what* quantity is plotted as well as to identify *who* created the study.

6. Click **[OK]** ✓ to accept these changes and close the **Stress Plot** property manager. A default plot of normal stress in the Y-direction (σ_y) now appears on the graphics screen and a new plot, named **Stress2 (-Y normal-)**, appears beneath the **Results** folder. The icon labeled **Stress1 (-vonMises-)** contains a plot of von Mises stresses while **Stress2** contains a plot of normal stress in the Y-direction (σ_y).

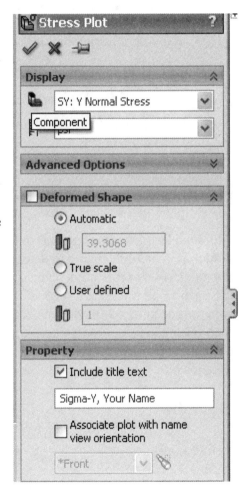

Figure 17 – Selections to specify plotting of a different stress component on the model.

Note that the title entered in step 5 appears at the top-left of the graphics window. Next, visual characteristics of the graphics display are modified to better view the results; to do this proceed as follows.

1-17

Analysis of Machine Elements using COSMOSWorks

7. Right-click **Stress2 (-Y normal-)** and from the pull-down menu, select **Settings…**. The **Settings** property manager opens as seen in Fig. 18.

8. Within the **Fringe Options** dialogue box, select **Discrete** from the pull-down menu. This action displays stress contours as discrete color bands (or fringes) rather than the rainbow effect created by the **Continuous** display. While on this menu, experiment with other **Fringe Options**, then reset to **Discrete** to correspond with images illustrated in this text.

9. In the **Boundary Options** dialogue box, select **Model**. This option outlines the model with a black line thereby making its edges easier to view.

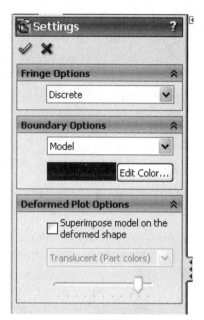

Figure 18 – Altering stress contour display options in the **Settings** property manager.

10. Click **[OK]** ✓ to close the **Settings** property manager. A *partial* image of the **Discrete** fringe plot of normal stress in the Y-direction appears in Fig. 19 (a).

Using the Probe Tool

Although stress contour plots provide a general sense of stress magnitudes throughout the model, it is often desirable to determine stress magnitudes at specific locations. To do this the **Probe** tool is used. To aid in selecting specific points on the model, two changes are introduced to create a second image similar to that of Fig. 19 (b). First, the **Mesh** is superimposed on the model and second, stress contours are represented by **Lines** rather than **Discrete** full-color fringes.

To alter model appearance, proceed as follows.

1. Return to the **Settings** property manager by right-clicking **Stress2 (-Y normal-)**. From the pull-down menu select **Settings…**.

Aside:
In this user guide the **Fringe options** field is specified as **Line** rather than **Discrete** because color shading makes the **Mesh** difficult to view when printed in black and white. Those who wish to retain the **Fringe Option** as **Discrete** are *encouraged* to do so and *skip* step 2 below.

---

2. Within the **Settings** property manager, change the **Fringe Options** field to **Line** *(Optional)*.

Basic Stress Analysis

3. Next, change the **Boundary Options** to **Mesh**.

4. Click **[OK]** ✓ to close the **Settings** property manager. The upper portion of the model should appear with the mesh shown as illustrated in Fig. 19 (b).

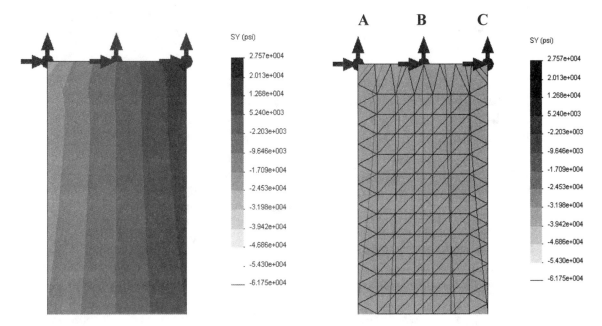

Figure 19 (a) – Upper portion of model showing stress σ_y as **Discrete** fringes.

Figure 19 (b) – Upper portion of the model showing normal stress σ_y using the **Line** option. A **Mesh** is superimposed on the model to facilitate using the **Probe** feature.

We are now ready to compare manually calculated stresses with stress values determined using finite element analysis. To facilitate comparisons, stress magnitudes at points A, B and C on top of the cam follower, Fig. 19 (b), are repeated below.

Stress at point A: $(\sigma_y)_A = \sigma_{Axial} + \sigma_{Bending} = (-2020) + (-22080) = -24,100$ psi

Stress at point B: $(\sigma_y)_B = \sigma_{Axial} + \sigma_{Bending} = (-2020) + (0) = -2,020$ psi

Stress at point C: $(\sigma_y)_C = \sigma_{Axial} + \sigma_{Bending} = (-2020) + (+22080) = +20,060$ psi

Begin by using the **Probe** tool to determine stresses at locations corresponding to points A, B, and C on the model and at all points in-between. It *may* be necessary to approximate the stress at the location of point B. The reason for this is that the automatically generated mesh size is somewhat arbitrary. Therefore, nodes, which typically are used as measurement points, may not lie *exactly* at the location of point B on the neutral axis. This will be determined as the analysis proceeds.

Begin by zooming-in on the top of the model. An image similar to that shown in Fig. 19 (a) or (b) should appear on your screen. Two different methods for using the Probe tool are demonstrated. Proceed as follows to use the **Probe** tool by the first method.

1-19

Analysis of Machine Elements using COSMOSWorks

5. In the COSMOSWorks manager tree right-click **Stress2 (-Y normal-)** and from the pull-down menu, select **Probe**. The **Probe Result** property manager opens as seen in Fig. 20. The table in the center of the **Results** dialogue box is *initially* empty.

6. In the **Options** dialogue box, click to choose ⊙ **On Selected Entities**. The **Results** dialogue box expands to include the highlighted (light blue) field. This field is active and waiting on the selection of **Faces, Edges or Vertices** on the model.

7. As the cursor is moved slowly over the model, the symbol adjacent to it changes to a square, a line, or a small circle to represent selection of a **Face**, an **Edge** or a **Vertex** respectively. Choose the line representing the top, front **Edge** of the cam follower. **Edge<1>** appears in the highlighted field and the **[Update]** button is active.

8. Clicking the **[Update]** button automatically fills in the **Results** table with the data described below.

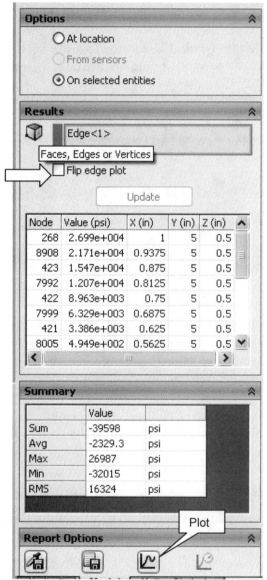

Figure 20 – **Probe Result** table containing stress values at node locations across the top **Edge** of the model.

*NOTE: To view complete results, it may be necessary to enlarge the table and individual columns within the table by clicking-and-dragging the right boundary of the **Probe Result** property manager and individual column boundaries respectively.*

Information contained in each column of the **Results** table includes (from left to right):

- **Node**: The number of each *sequential* node located across the top front edge of the model. Node numbers are assigned automatically by the software during mesh generation. Recall that there are 16 nodes across the top edge because high quality tetrahedral elements have nodes at both mid-side and at corner locations on each element. [Count element sides in Fig. 19 (b)].

1-20

Basic Stress Analysis

- **Value (psi)**: Magnitude of the *selected* stress at each node location. For the current example, the **Value (psi)** column contains magnitude of stress in the Y-direction, (σ_y).

- **X (in)**: X-coordinate of each node location. The initial value appears as 1 inch and subsequent values decrease in 1/16 inch increments across the model.

- **Y (in)**: Y-coordinate of each node location. The original SolidWorks model of the cam follower was created by locating its bottom left-hand corner at the origin of a global coordinate system. Thus, all points along the top edge lie 5 inches above the origin.

- **Z (in)**: Z-coordinate of each node location. Because the original model was extruded ½ inch in the positive Z direction, all Z-coordinates on the top front edge are located at Z = + 0.5 inches.

9. In the **Report Options** dialogue box, located at the bottom of the **Probe Result** property manager, click the **Plot** icon; see flag in Fig. 20. Immediately a graph of stress σ_y across the top edge of the model is displayed.

Notice, however, that this graph displays tensile (positive) stress on its left-hand side and compressive (negative) stress on its right-side. This is just the reverse of the actual stress distribution across the top of the cam follower. The reason for this is that the top edge is a line with two different ends, but no means is provided to select one end or the other. This logical misrepresentation of data is easily remedied by the following action.

10. First, close the current graph by clicking the ☒ at its upper-right corner.

11. Next, in the **Results** dialogue box, place a check ✓ adjacent to ☑ **Flip edge plot**. See arrow in Fig. 20.

12. Again, click the **Plot** icon and a corrected graph, showing variation of σ_y across the top edge of the model, from left to right, appears in Fig. 21.

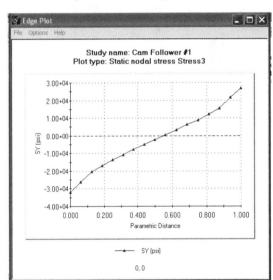

Figure 21 – Graph of stress distribution σ_y from the left side (0.0 in.) to right side (at 1 in.) along top front edge of the cam follower.

The graph in Fig. 21 indicates compressive stress along the left side of the model as expected. This stress gradually decreases to zero near the center of the model and then becomes positive, indicating tensile stress, on the right side of the cam follower. These results are consistent with our understanding of stress distribution on the top edge of the model. However, we next observe that these results do not agree well with classical theory.

OBSERVATIONS:
Table 1 compares results of manually calculated stresses with finite element results obtained using the **Probe** tool at locations A, B, and C at the top-end of the cam follower. Verify results in the Probe Tool Results column of Table 1 by scrolling through values in the **Results** table.

Table 1 – Comparison of stress σ_y computed by classical and finite element methods.

Location	Manual Calculation (psi)	Probe Tool Results (psi)	Percent Difference (%)
Point A	-24100	-32020	24.7 %
Point B	-2020	-2224	9.1 %
Point C	+20060	+26990	25.7 %

Results in Table 1 appear to indicate significant differences between results calculated using classical stress equations and those determined by finite element analysis methods. How can this be?

The above question can be answered by recalling St. Venant's principle, which states that stress predicted by classic equations exists only in regions *reasonably well removed* from points of load application, support locations, or locations of geometric discontinuity. Any of these conditions typically introduce significant *localized* effects. Thus, in these regions, shortcomings of classical stress equations are at odds with the enhanced predictive capability of finite element results. For this reason, the **Probe** tool is used again, but this time it is used to select nodes at locations somewhat removed from the upper support. Proceed as follows.

13. Begin by closing the **Edge Plot** graph. Click ☒ to close the graph.

14. Within the **Options** dialogue box, click to select ⦿ **At location**. This action clears all data from the **Results** table and changes the mode of operation to one that enables display of results at user selected node locations.

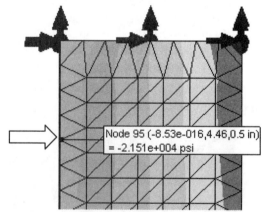

Figure 22 – Close-up view of top end of the model showing the first node selected along an *approximate* straight-line across the model.

15. Move the Probe tool into the graphics area and count down to the 4th node below the top-left corner. See node selected in Fig. 22. This node is somewhat

Basic Stress Analysis

arbitrarily selected to be "sufficiently removed" from the *immovable* **Restraint** condition at the top of the cam follower. This is done to avoid the effects predicted by St. Venant's principle.

As you perform the next step, a small text box will appear adjacent to each selected node. Information in this box lists the *Node Number* followed, in parenthesis, by its *(X, Y, Z)* coordinates. The second line of text lists the stress magnitude at the selected node; units are included with all values. These values are also listed in the **Results** table of the **Probe Result** property manager. Unfortunately, the text-boxes tend overwrite one another for closely spaced nodes. However, when used to identify stress at specific locations on a model, they can be quite useful. Rotate the model as shown in Fig. 23 to facilitate selecting nodes that otherwise would be hidden behind these text boxes.

16. Proceed across the model *from left-to-right* and click to select *only* nodes at element corners as illustrated by darkened dots in Fig. 23. If an error is made when selecting nodes, simply click the ⊙ **At Location** button again. This action clears the **Results** table and the selection process can be repeated.

Aside:
Due to the arbitrary mesh generation scheme it may not be possible to select a row of nodes in a *straight line* across the model. Despite this inconvenience, it is still possible to compute stresses at each node location and compare values obtained using classical stress equations with finite element results listed in the **Results** table.

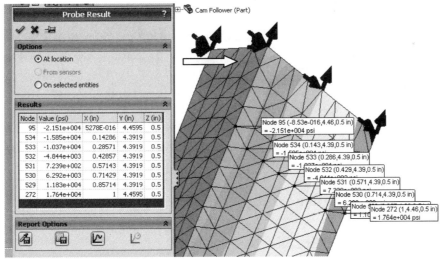

Figure 23 – **Results** table containing σ_y values and node locations for node selection across the model at a distance "sufficiently removed" from support conditions.

17. Within the **Report Options** dialogue box, once again select the Plot icon. The resulting **Probe Result** graph is shown in Fig. 24.

The graph in Fig. 24 reveals the expected variation of stress caused by adding the uniform axial compressive stress to the near linear variation of bending stress. Bending

1-23

stress varies from compression on left side of the model to tension on the right side. Also in Fig. 24, note the cross-hairs (intersecting dashed lines), located at the centroidal axis of the model correspond to the middle node (X = 3.5 nodes) and stress value (σ_y = 2014.93 psi) at this location. Compare this stress value with those listed at point B in Table 2 below.

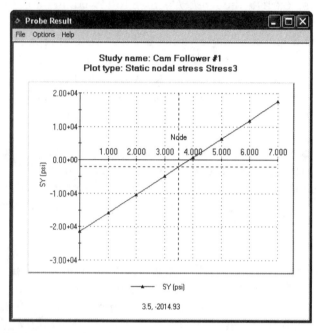

Figure 24 – Plot of combined axial and bending stress variation (σ_y) across the model.

Table 2 shows a comparison between stresses calculated using classical equations and those computed using the finite element analysis. Results of both methods are compared at *approximately* 0.6 inches below the top of the model. Refer to the Y-coordinate in the **Results** table of Fig. 23 for the exact distance (5.000 in. - 4.3919 in. = 0.6081 in.) for most nodes.

Table 2 – Comparison of stress σ_y computed by classical and finite element methods at an arbitrary distance below the top of the cam follower.

Location	Manual Calculation (psi)	Probe Tool Results (psi)	Percent Difference (%)
Point A	-21710	-21510	0.92 %
Point B	-2020	-2015	0.25 %
Point C	17670	17640	0.17 %

A significant improvement in the comparison of computed results is observed in Table 2. Differences are less than 0.92% at all points A, B and C. Based on these comparisons, several observations are included in the Summary section below.

18. Click ☒ to close the **Probe Result** graph.

19. Click **[OK]** ✓ to close the **Probe Result** property manager.

Summary

- The effect of *localized* conditions, also known as "end-conditions" or "boundary conditions" (i.e., at support locations and locations of applied loads), upon finite element results can be significant. These effects are not predicted by classical equations used in manual stress calculations. For this reason care must be exercised when checking finite element results in these regions.

- Excellent agreement typically does exist between finite element analysis and manually computed results at locations *sufficiently removed* from localized conditions (examples of St. Venant's principle in operation).

- The importance of selecting the *appropriate stress* within a finite element analysis when comparing results cannot be overstated. For example, the COSMOSWorks default plot of von Mises stress does not, and should not, agree if compared with manually calculated values of stress in the Y-direction (σ_y). Simply stated, *like* stresses must be compared when checking validity of a finite element analysis.

The current example concludes at this point. Most of the software capabilities introduced in this example are fundamental to a successful finite element analysis and will be encountered repeatedly in subsequent examples and end-of-chapter problems.

This file either can be saved or deleted. A brief description of the two options for closing files follows. Because a variety of file structures are found in different computer work environments, the guidelines below are quite general. It is suggested that local system guidelines be followed regarding *where* files are saved (i.e., to a personal USB or "jump" drive, to the hard-drive or to personal file space allocated on a system network).

CLOSE the file and SAVE it to your account:
To exit COSMOSWorks and save a file *to your file space*, select **File** from the main menu. Next select **Save As...** this opens the **Save As** window. When queried, **Save results**; click **Yes** to save the results calculated during the solution process. Since work is to be saved to your personal account space, follow the file handling protocol for your computer system or network. Then close **SolidWorks** as you would any other Windows program.

CLOSE the file WITHOUT SAVING:
To exit COSMOSWorks *without saving your results*, select **File** from the main menu. From the pull-down menu, select **Close**. The **SolidWorks** window opens and displays the message "**Save changes to Cam Follower (Part)?**" choose **[No]**. Close **SolidWorks** as you would any other Windows program.

EXERCISES

End of chapter exercises are intended to provide additional practice using principles introduced in the current chapter. Future chapters also build upon capabilities mastered in preceding chapters. SolidWorks part files for all example and end-of-chapter problems can be downloaded from: http://www.schroff1.com/resources

Most exercises include multiple parts. Maximum benefit is realized by working all parts. However, in an academic setting, it is likely that parts of problems will be assigned or modified to suit specific course goals.

> **Default Expectations (unless specified otherwise)**
> - Include a *descriptive* Study name and *your name* when naming each Study. This approach uniquely labels each printed page with your name.
> - Use *discrete* stress contours (colored fringes) and display all plots on an *un-deformed* view of the model.
>
> RECALL: "Plot" – refers to a stress contour plot (i.e., colored fringes depicting stress magnitudes within a model).
> "Graph" – refers to a X,Y line-graph that depicts a relationship between two variables.

1. The cantilever beam pictured in Fig. E1-1 is rigidly supported (**Immovable**) at its left-end and is subject to forces F_x and F_y applied to its right-end. Either create a beam model using SolidWorks on your own or open problem file **Cantilever 1-1**, which is available at the publisher's web site (see above). Then, perform a finite element analysis of this beam subject to the following guidelines.

    - Material: **AISI 1045 Steel cold drawn**

    - Mesh: **High Quality** tetrahedral elements (system default)

    - Assign an axial **Force** F_x = 830 lb that acts **Normal** to the right-end of the beam in the direction shown. Also apply a downward **Force** F_y = 760 lb that acts on the top-right **Edge** of the beam. *Split Lines* are not used.

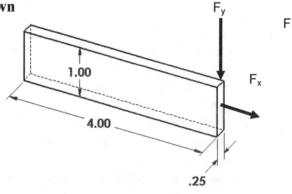

Figure E1-1 – Basic dimensions of a cantilever beam subject to loads F_x and F_y.

    - Assign an **Immovable** restraint (not shown above) to the left-end of the beam

Basic Stress Analysis

Determine the following:
  a. Create a stress contour plot of shear stress τ_{xy} showing the stress distribution on the front face, which corresponds to the 1.00 x 4.00 in side of the beam. Include your name and a descriptive title on this plot.

  b. Use the **Probe** feature to create a graph of τ_{xy} across the left-front edge of the beam. Create this graph by zooming in on the left end of the model and selecting its left edge; use **Flip edge plot** if necessary.

  c. Use the **Probe** feature to create a graph of τ_{xy} across the front face of the beam by selecting successive nodes (from top to bottom) at a location approximately 1-in. to the right of the fixed-end of the model. (Refer to the **Results** table to ensure a X-value corresponding to approximately 1 in. from the left-end of the model is used). If nodes are not aligned in a straight line across the model, then choose nodes along the best approximation of a straight line. On this plot, manually label the value and location of τ_{xy} used to determine maximum shear due to bending.

  d. Calculate the maximum shear stress due to bending using classical stress equations at the fixed end and 1 in to the right of the fixed end. Then, use equation [1] below to compute the percent difference between classical and finite element solutions for τ determined at the center of the beam (between top and bottom surfaces) in parts (b) and (c) above.

  $$\% \text{ difference} = \frac{(\text{FEA result - classical result})}{\text{FEA result}} * 100 = \quad\quad [1]$$

  e. Briefly state reason(s) why shear stress determined in parts (b) and (c) differ.

  f. Use the **Probe** feature to create a graph of σ_x across the front face of the beam at approximately 1-in. to the right of the fixed-end. Select *both* corner and mid-side nodes to yield a more uniform graph. In the **Results** table, observe and record the exact X-coordinate of nodes located on the top and bottom edges of the beam. Assign a descriptive title and axis labels to this graph.

  g. Use classical stress equations to calculate the *appropriate* normal stress on *both* the top and bottom surfaces of the beam at approximately 1-in. to the right of the fixed-end. Use the "correct distance" to calculate stress at this location. Hint: the X-coordinate of node location(s), determined in step (f) above, should help to determine the "correct distance."

  h. Use equation [1] above to compare finite element results for σ_x with manual calculations of σ_x on both top and bottom surfaces of the beam. Clearly label calculations corresponding to "top" and "bottom" surfaces.

  i. If results of parts (d) or (h) differ by more than 2% locate and correct the error(s) in either the finite element model and/or in manual stress calculations.

2. The beam pictured below is rigidly supported **(Immovable)** at its left-end and simply supported near its right end. It is made from **Alloy Steel** and is subject to a concentrated load of F = 1 kN at the location shown. Open the file **Beam 1-2**. Alternatively, this simple part may be created on your own. However, users unfamiliar with creating a *Split Line* should use the model provided at the publisher's web site (http://www.schroff1.com/). Create a finite element model of this beam assuming the following.

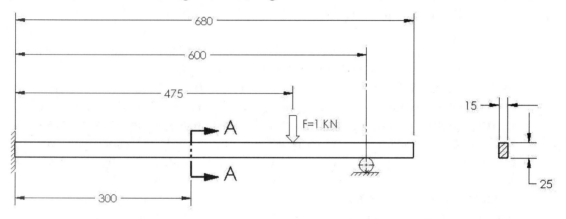

Figure E2-1 – Rectangular beam "fixed" at its left-end and supported by a roller near its right end. The applied load F = 1 kN. All lengths in mm.

- Material: **Alloy Steel (SS)** (Use S.I. units)

- Mesh: **High Quality** tetrahedral elements

- Assign the applied **Force** downward as shown on the top *Split Line* and **Immovable** restraints at both the left-end *and* at the roller support.

Determine the following:
a. Use classical mechanics of materials equations to compute bending stress and shear due to bending on the top, bottom, and centroidal axis of the beam at section A-A.

b. Create a stress contour plot of bending stress throughout the beam.

c. Use the **Probe** feature to create a graph of bending stress at location A-A. Create this graph by zooming in on the model and selecting node points, from top to bottom, across the model. Because a **High** quality mesh quality is used, be sure to select all corner *and* mid-side nodes *in sequence*. Use equation [1] (repeated below) to compute the percent difference between stresses predicted by classical equations in part (a) and the finite element solution at top, centroidal (middle), and bottom surfaces of the beam.

$$\% \text{ difference} = \frac{(\text{FEA result - classical result})}{\text{FEA result}} * 100 = \qquad [1]$$

Basic Stress Analysis

d. Use the **Probe** feature to create a graph of shear stress due to bending across the front face of the beam at location A-A. Select successive nodes, from top to bottom of the beam, and compute the percent difference between τ at the neutral axis determined by both classical and finite element methods.

e. Repeat parts (a) and (d) of the problem, but this time use a different restraint to model the roller support as outlined in the following steps.

- Right-click the **Restraint-n** folder, where "**n**" corresponds to the restraint applied at the roller support. Then, from the pull-down menu, select **Edit Definition...**. The **Restraint** property manager opens. See Fig. E2-2.

- From the pull-down menu at the top of the **Type** dialogue box, select **Use Reference Geometry**.

- In the field beneath the pull-down menu (the **Faces, Edges, Vertices for Restraint** field), the *Split Line* at the roller support should appear as **Edge<1>** (provided the proper **Restraint-n** folder was opened in the first step). The restraint applied to this line is edited in the following steps.

- In the second field beneath the pull-down menu in the **Type** dialogue box, select the bottom face of the beam as the **Face, Edge, Plane, Axis for Direction**. **Face<1>** should appear in this field as illustrated in Fig. E2-2.

- Click ⌄ to open the **Translations** dialogue box.

- Although it makes little difference in what follows, set **Unit** to **mm**.

Figure E2-2 – Editing the **Restraint** property manager to specify a different restraint at the roller support.

- At the bottom of the **Translations** dialogue box, click the **Normal to plane** icon, see arrow in Fig. E2-2, and enter "**0**" (if not already present) to specify zero displacement is allowed normal to the line on the bottom surface.

1-29

Analysis of Machine Elements using COSMOSWorks

- Click **[OK]** ✓ to close the **Restraint** property manager.

- Click the file name initially specified for this analysis and then select **Run** to execute a new solution to the problem using the modified restraint at the roller support.

f. Describe which of the boundary conditions (i.e., **Restraints**) applied at the roller support produces the best approximation to classical results. Compare such items as percent difference between classical and finite element results to substantiate answers. Also, describe why one set of **Restraints** is a better model of physical reality than the other.

Textbook Problems
It is highly recommended that the above exercises be supplemented by problems from a design of machine elements textbook. A great way to discover errors made in formulating a finite element analysis is to work problems for which the solution is known by independent calculation or experiment. Typical textbook problems, if well defined in advance, make an excellent source of solutions for comparison.

CHAPTER #2

CURVED BEAM ANALYSIS

This example, unlike the first, will lead you quickly through those aspects of creating a finite element Study with which you already have experience. However, where new information or procedures are introduced, additional details are included. For consistency throughout this text, a common approach is used for the solution of all problems.

Learning Objectives
In addition to software capabilities studied in the previous chapter, upon completion of this example, users should be able to:

- Use COSMOSWorks *icons* in addition to menu selections.

- Apply a *split line* to divide a selected face into one or more separate faces.

- Simulate *pin loading* inside a hole.

- Use **Design Checks** to determine the *safety factor* or lack thereof.

- Determine *reaction* forces acting on a finite element model.

Problem Statement
A dimensioned model of a curved beam is shown in Fig. 1. Assume the beam material is 2014 Aluminum alloy, and it is subject to a downward vertical force, $F_y = 3800$ lb, applied through a cylindrical pin (not shown) in a hole near its free end. The bottom of the curved beam is considered "fixed." In this context, the *actual* fixed end-condition is analogous to that at the end of a cantilever beam where translations in the X, Y, Z directions and rotations about the X, Y, Z axes are considered to be zero. However, recall from Chapter 1 that **Restraint** types within COSMOSWorks also depend on the type of element to which they are applied. Therefore, because solid tetrahedral elements are used to model this curved beam **Immovable** restraints are used.

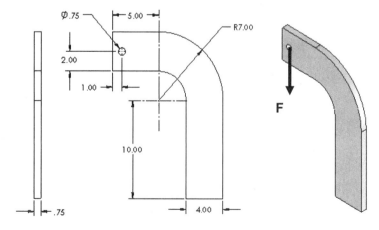

Figure 1 – Three dimensional model of curved beam.

Design Insight

Numerous mechanical elements occur in the shape of initially curved beams. Examples include: C-clamps, punch-press frames, and bicycle caliper brakes, to name a few. This example examines the stress at section A-A shown in Fig. 2. Section A-A is chosen because it is the furthest distance from the applied force **F** thereby creating reaction force **R = F** and the maximum bending moment **M = F*L** at that location. Accordingly, classical equations for stress in a curved beam predict maximum stress at section A-A.

In this example, the validity of this common assumption is investigated while exploring additional capabilities of the COSMOSWorks software listed above.

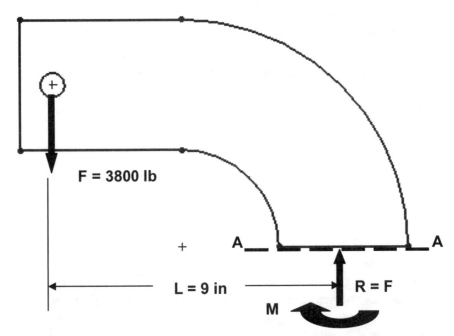

Figure 2 – Traditional free-body diagram of the upper portion of a curved beam model showing applied force **F** acting at a hole, and reactions **R = F**, and moment **M** acting on section **A-A**.

Creating a Static Analysis (Study)

1. Open SOLIDWorks by making the following selections. These steps are similar to opening any Windows program. (*Note:* A "/" is used below to separate successive menu selection)

    Start / All Programs / SolidWorks 2008

2. In the SolidWorks main menu, select **File / Open…** Then browse to the location where COSMOSWorks files are stored and open the file named "**Curved Beam.**"

If a pop-up COSMOSWorks window appears and states: **The following documents will be converted when saved "Curved Beam.SLDPRT."** Click **[OK]** to close the window.

> **Reminder:**
> If you do not see **COSMOSWorks** listed in the main menu of the SolidWorks screen, click **Tools/Add-Ins...**, then in the **Add-Ins** window check ☑ **COSMOSWorks 2008** in both the **Active Add-Ins** and the **Start Up** columns, then click **[OK]**. This action adds the **COSMOSWorks** icon to the top of the manager tree and to the main menu.

3. Click the COSMOSWorks analysis manager icon located at the top right-side of the manager tree shown in Fig. 3. This action switches the user into COSMOSWorks.

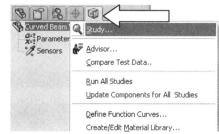

Figure 3 – Opening a Study in COSMOSWorks.

Because another goal of this analysis is to introduce use of COSMOSWorks icons, verify that the icons shown in Fig. 4 appear in the toolbar on your screen. If they do not appear, proceed as follows, otherwise skip to step 6.

Figure 4 – Standard COSMOSWorks icons.

4. Right-click anywhere in the toolbar at top of the screen. This action opens the COSMOSWorks Command Manager menu shown in Fig. 5.

5. Within the Command Manager menu, select:
   COSMOSWorks Loads
   COSMOSWorks Main
   COSMOSWorks Result Tools
   The above selections are shown "boxed" in Fig. 5.

Toolbars may be displayed in different locations on the screen than those shown in Fig. 4. Reposition the toolbars by clicking-and-dragging the "handles" located at the left-end of each toolbar.

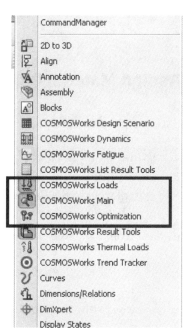

Figure 5 – Opening standard COSMOSWorks toolbars.

2-3

Analysis of Machine Elements using COSMOSWorks

6. In the COSMOSWorks manager, Fig. 3, right-click **Curved Beam** and from the pull-down menu select **Study…** Alternatively, from the COSMOSWorks toolbar, select the **Study** icon . The **Study** property manager opens (not shown).

7. In the **Name** dialogue box, replace **Study 1** by typing a descriptive name. For this example, type: **Curved Beam Analysis-YOUR NAME**. Recall that including your name along with the Study name ensures that it is displayed on each plot.

8. Examine other fields within the **Study** property manager to ensure that a **Solid Mesh** is specified and that **Static** is selected as the analysis **Type**.

9. Click **[OK]** ✓ (green check mark) to close the **Study** property manager.

The COSMOSWorks manager, seen in Fig. 6, is updated to show an outline for the current Study.

As in the previous example, the sequence of steps outlined in Fig. 6 is followed from top to bottom as the current finite element analysis is developed.

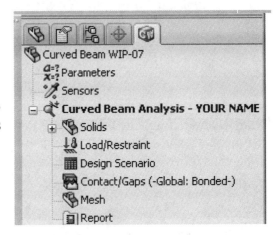

Figure 6 – Basic steps of a Study.

Assign Material Properties to the Model

Part *material* is defined as described below.

1. In the COSMOSWorks manager tree, right-click the **Solids** folder and from the pull-down menu select **Apply Material to All…** Alternatively, click the "+" sign adjacent to the **Solids** folder followed by clicking the folder labeled **Curved Beam**. Then, in the newly added toolbar, select the **Apply Material to Selected Components** icon . The **Material** window opens as shown in Fig. 7.

   a. Under **Select material source**, click to select ⦿ **From library files**.

   b. From the pull-down menu, select **cosmos materials** (if not already selected).

2-4

c. Click the plus sign next to the **Aluminum Alloys (85)** and scroll down to select **2014 Alloy**. The properties of 2014 Aluminum alloy are displayed in the right-half of the window.

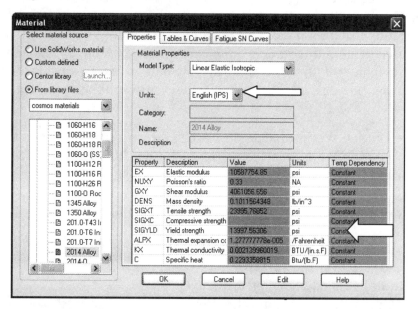

Figure 7 – Material properties are selected and/or defined in the **Material** window.

d. In the right-half of this window, select the **Properties** tab (if not already selected). Change **Units:** to **English (IPS)**.

Note the material Yield strength denoted by **SIGYLD Yield Strength** is a relatively low 13997.563 psi (approximately 14000 psi). Material with a low yield strength is intentionally chosen to facilitate discussion of *Safety Factor* later in this example. Examine other values in the table to become familiar with data available in the material properties library.

e. Click **[OK]** to close the **Material** window. A check "✓" mark appears on the **Solids** folder to indicate a material has been selected.

Aside:
If at any point you wish to *change the material specification* of a part, such as during a redesign, right-click the **Solids** icon in the manager tree and select **Apply Material to All…**. The **Material** window is opened and an alternative material can be selected or defined.

However, be aware that when a different material is specified *after* running a solution, it is necessary to run the solution again using the revised material properties.

Applying Restraints

For a static analysis, adequate restraints must be applied to stabilize the model. In this example, the bottom surface of the model is considered "fixed." However, recall when defining this **Study** that a *solid* mesh was specified. Therefore, because tetrahedral elements are used to mesh this model, "Immovable" restraints are applied to the bottom surface. Bear in mind that "fixed" restraints apply only to *Shell* elements. Shell elements are introduced in a later chapter. Proceed as follows to specify this restraint.

1. In the COSMOSWorks toolbar left-click the **Restraints** icon. Alternatively, in the COSMOSWorks manager, right-click **Load/Restraint** and select **Restraints...**

   The **Restraint** property manager opens as shown in Fig. 8.

2. Under **Type**, click to open the pull-down menu and select **Immovable (No translation)**. This designation restricts translations in X, Y, and Z directions only.

3. The **Faces, Edges, Vertices for Restraint** field immediately below the pull-down menu is highlighted (light blue) to indicate it is active and waiting for the user to select part of the model to be restrained. Rotate, and/or zoom to view the bottom of the model. Next, move the cursor over the model and when the bottom surface is indicated, click to select it. The surface is highlighted and, provided the ☑ **Show preview** box is checked, restraint symbols appear as shown in Fig. 9. Also, **Face<1>** appears in the **Faces, Edges, Vertices for Restraint** field.

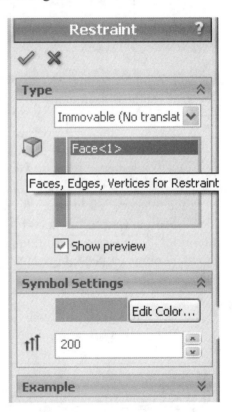

Figure 8 – **Restraint** property manager.

If an incorrect entity (such as a vertex, edge, or the wrong surface) is selected, right-click the incorrect item in the **Faces, Edges, Vertices for Restraint** field and select **Delete**, then repeat step 3.

4. Click the down arrow to open the **Symbol settings** dialogue box.

5. Both color and size of the restraint symbols (vectors in the X, Y and Z directions) can be changed by altering values in the **Symbol Settings** field of Fig. 8. Experiment by clicking the up ▲ or down ▼ arrows to change size of restraint symbols. A box of this type, where values can be changed either by typing a new value or by clicking the ▲▼ arrows, is called a "spin box." Restraint symbols shown in Fig. 9 were arbitrarily increased to 200%. Experiment with this option.

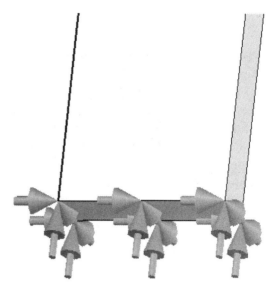

Figure 9 – **Immovable** restraints applied to bottom of the curved beam model.

6. Click **[OK]** ✓ (green check mark) at top of the **Restraint** property manager to accept this restraint. An icon named **Restraint-1** appears beneath the **Load/Restraint** folder in the COSMOSWorks manager.

Aside:
Immovable restraint symbols shown in Fig. 9 appear as simple arrows or vectors with a straight shaft. These symbols indicate that *displacements* in the X, Y and Z directions are restrained (i.e., prevented).

Fixed restraints are introduced in Chapter 4 where a small disk is added to the tail of each restraint symbol. This difference of symbols indicates that *rotations* about the X, Y and Z axes are also restrained. Watch for this subtle difference in future examples.

Applying Load(s)

Next apply the downward force, $F_y = 3800$ lb, at the hole located near the top left-hand side of the model shown in Figs. 1 and 2. This force is assumed to be applied by a pin (not shown) that acts through the hole.

Analysis Insight

Because the goal of this analysis is to focus on curved beam stresses at Section A-A, and because Section A-A is well removed from the point of load application, modeling of the applied force can be handled in a number of different ways. For example, the downward force could be applied to the vertical *surface* located on the upper-left side of the model, Fig. (a). Alternatively, the force could be applied to the upper or lower *edge* of the model at the extreme left side, Fig. (b). These loading situations would require a slight reduction of the magnitude of force **F** to account for its additional distance from the left-side of the model to section A-A (i.e., the moment about section

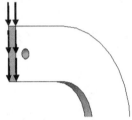 

A-A must remain the same).

Figure (a) – Force applied to left surface.   Figure (b) – Force applied to lower edge.

The above loads are simple to apply. However, the assumption of pin loading allows us to investigate use of a *Split Line* to isolate a *portion* of the bottom of the hole surface where contact with a pin is assumed to occur. This surface is where a pin force would be transferred to the curved beam model. The actual contact area depends on a number of factors, which include geometries of the contacting parts (i.e., relative diameters of the pin and hole) and material properties (i.e., hard versus soft contact surfaces of either the pin or the beam). This example arbitrarily assumes a reasonable contact area so that use of a *Split Line* can be demonstrated. If, on the other hand, contact stresses in the vicinity of the hole were of paramount importance, then determination of the true contact area requires use of *Contact/Gap* analysis. This form of analysis is investigated in Chapter 6.

Inserting Split Lines

The first task is to isolate a portion of the area at the bottom of the hole. This can be accomplished by using a *Split Line*. The method described below outlines the use of a reference plane to insert a *Split Line*.

1. From the main menu, select **Insert**. Then, from succeeding pull-down menus make the following selections: **Reference Geometry** followed by **Plane…**. The **Plane** property manager opens *and* the SolidWorks "flyout" menu also appears at the upper-left corner of the graphics screen as shown in Fig. 10.

2. On the graphics screen it may be necessary to click the "+" sign adjacent to the SolidWorks **Curved Beam** icon to display the complete "flyout" menu as illustrated in Fig. 10. Continue to refer to Fig. 10 in steps 3 through 5 below.

3. Within the **Plane** property manager, under **Selections**, the **Reference Entities** field is highlighted (light blue). Note text appearing in the *Status Line* at bottom left of the screen, which prompts the user to "**Select valid entities to define a plane (plane, face, edge, line or point)**." In the following steps, a horizontal reference plane that passes through the bottom portion of the hole is defined.

4. From the SolidWorks flyout menu, select **Top Plane**. For users who opened the **Curved Beam** part file available at the textbook web site, the top plane passes through the part origin, which is located on the bottom of the model[1] in Fig. 10.

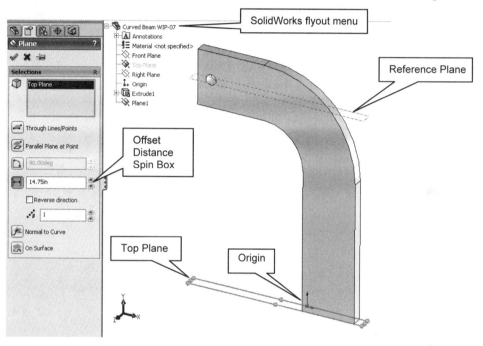

Figure 10 – The **Plane** property manager and various selections made to create a reference plane that passes through the bottom of the hole.

5. In the **Plane** property manager, return to the **Selections** dialogue box and in the **Offset Distance** spin-box, type **14.75**. This is the distance *from* the **Top Plane** to a reference plane located so that it passes through the bottom portion of the hole.

Aside:
The 14.75 in. dimension is determined from the following calculation. Refer to Fig. 1 to determine the source of values used in the equation below.

10 in (height of straight vertical sides) + 3 in (radius of concave surface) + 1.75 in (distance from the horizontal edge beneath the hole and extending into the bottom portion of the hole) = 14.75in.

It is emphasized that the area intersected on the bottom of the hole is chosen *arbitrarily* in this example!

---

[1] Users who created a curved beam model from scratch can also follow these instructions. The *only* difference being specification of the proper *distance* from the **Top Plane** (used as a reference in *your* model) to the bottom of the hole.

Analysis of Machine Elements using COSMOSWorks

6. Click **[OK]** ✓ to close the **Plane** property manager.

The reference plane created in the preceding steps appears highlighted in the current screen image. In the following steps this plane is used to create *Split Lines* near the bottom of the hole. These *Split Lines* enable us to define a small area on the bottom of the hole where the downward load will be applied.

7. From the main menu, select **Insert**. Then, from the pull-down menus choose: **Curve** followed by **Split Line…** The **Split Line** property manager opens as shown in Fig. 11.

8. Beneath **Type of Split**, select ⦿ **Intersection**. This choice designates the means by which *Split Lines* are defined for this example (i.e., they will be located where **Plane1** *intersects* the hole).

9. In the **Selections** dialogue box, **Plane1** may already appear in the **Splitting Bodies/Faces/Planes** field. If **Plane1** does not appear in this field, click to activate the field (light blue), then move the cursor onto the graphics screen and select the upper plane when it is highlighted. **Plane1** now appears in the top field.

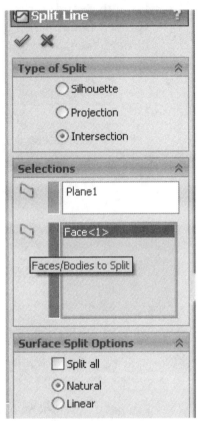

Figure 11 – **Split Line** property manager showing selections.

10. Next, click inside the **Faces/Bodies to Split** field. This field may already be active (light blue). Then move the cursor over the model and select anywhere on the *inside surface* of the hole. It may be necessary to zoom in on the model to select this surface. Once selected, **Face<1>** appears in the active field. Figure 12 shows a partial image of the model with *Split Lines* appearing where **Plane1** intersects the bottom of the hole.

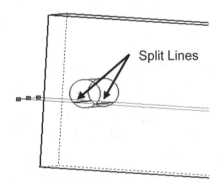

Figure 12 – Close-up view of hole showing *Split Lines* where **Plane1** intersects near the bottom of hole.

Remain zoomed-in on the model to facilitate applying a force to the inside of the hole.

11. In the **Surface Split Options** dialogue box, select ⊙ **Natural**. A **Natural** split follows the contour of the selected model surface.

12. Click **[OK]** ✓ to close the **Split Line** property manager.

13. If an information "flag" appears adjacent to the *Split Lines*, click ⊠ to close it.

Applying Force to an Area Bounded by Split Lines

Now that a restricted area on the bottom of the hole has been identified, the next step is to apply a downward force, $F_y = 3800$ lb, on this area. Proceed as follows.

Figure 13 – Specifying a force and its direction on the hole bottom.

1. In the COSMOSWorks toolbar, select the **Force** icon. Alternatively, in the COSMOSWorks manager, right-click the **Load/Restraint** icon and from the pop-up menu select **Force…**. A partial view of the **Force** property manager appears in Fig. 13.

2. Under **Type**, click to select ⊙ **Apply force/moment**. Again the **Faces, Edges, Vertices, Reference Points for Force** (light blue) field prompts the user to select an appropriate entity.

3. Move the cursor over the model and when the *bottom inside surface* of the hole is outlined, click to select it. **Face<1>** appears in the active field of the **Type** dialogue box.

4. Next, click to activate the second field from the top of the **Type** dialogue box. Passing the cursor over this field identifies it as the **Face, Edge, Plane, Axis for Direction** field. This field is used to specify the direction of the force applied to the bottom of the hole.

   Because a downward, vertical force is to be applied, select a *vertical* edge . . . *any vertical edge* . . . on the model aligned with the Y-direction. After selecting a vertical edge, **Edge<1>** appears in the active field and force vectors appear on the model as seen in Fig. 14.

5. Set **Units** to **English (IPS)**, if not already selected.

2-11

6. In the **Force (Per entity)** field, type **3800**. As noted in an earlier example, it may be necessary to check ☑ **Reverse Direction** if the force is not directed *downward*.

7. Click **[OK]** ✓ to accept this force definition and close the **Force** property manager. An icon named **Force-1** appears beneath the **Load/Restraint** folder of the COSMOSWorks manager.

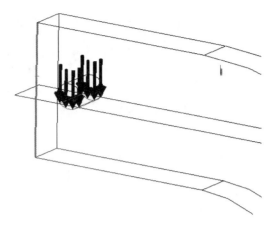

Figure 14 – Downward force applied between *Split Lines* on bottom of hole. A wireframe view of model is shown.

The model is now complete as far as material, restraint, and force definitions are concerned. The next step is to mesh the model as described below.

Meshing the Model

1. In the COSMOSWorks toolbar, select the **Mesh** icon. Alternatively, in the COSMOSWorks manager, right-click **Mesh** and from the pull-down menu select **Create Mesh...** The **Mesh** property manager opens as shown in Figs. 15 (a & b).

2. In the **Mesh Parameters:** dialogue box, set **Unit** to **in** (if not already selected). Accept remaining default settings (i.e. mesh size and tolerance) in this dialogue box. These system-determined values usually provide sufficient definition for an initial analysis on a non-complex part.

3. Click the down arrow to open the **Options** dialogue box. Most of the following selections should appear as default settings. These settings should produce a reasonably good quality mesh. However, verify that the settings are as listed below and only change them if they differ.

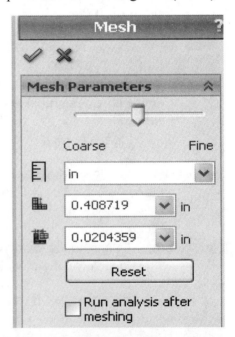

Figure 15 (a) – **Mesh** property manager showing system default **Mesh** settings applied to the current model.

4. Under **Quality**, select ⊙ **High**.

5. Under **Mesher:**, select ⊙ **Standard**.

6. Under **Mesher options:** check ☑ **Jacobian check for solid:** and the pull-down menu should be set to **4 points**.

7. Click the **[Apply]** button to place these settings into effect.

8. Finally, click **[OK]** ✓ to accept these settings and close the **Mesh** property manager.

Meshing starts automatically and the **Mesh Progress** window appears briefly. After meshing is complete, COSMOSWorks displays the meshed model shown in Fig. 16. Also, a check mark "✓" appears on the **Mesh** folder to indicate meshing is complete.

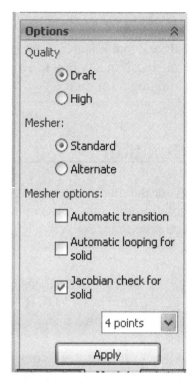

Figure 15 (b) – View of the **Options** portion of the **Mesh** property manager.

OPTIONAL:
9. Display mesh information by right-clicking the **Mesh** folder and select **Details...**

The **Mesh Details** window displays a variety of mesh information. Scroll down the list of information an note the number of nodes and elements for this model is 12554 nodes and 7213 elements (numbers may vary slightly due to the automated mesh generation procedure).

Rotate the model as illustrated in Fig. 16 and notice that the mesh is just two elements *thick*. Two elements across the model's thinnest dimension are considered the minimum number for which *Solid Elements* should be used. Thus, two elements are considered an unofficial dividing line between when *Shell* or *Solid Elements* should be used. Therefore, either element type could be used for this model. But keep in mind that shell elements are typically reserved for thin parts.

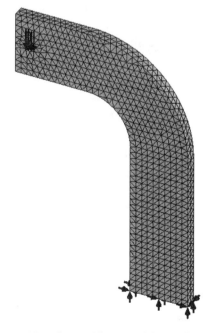

Figure 16 – Curved beam with mesh and boundary conditions illustrated.

Analysis of Machine Elements using COSMOSWorks

> **Reminder**
> Recall that it is permissible to define material properties, restraints, forces, and create the mesh in any order. However, all these necessary steps must be completed before running a solution.

Solution

After the model has been completely defined, the solution process is initiated. During a solution the numerous equations defining a Study are solved and results of the analysis are saved for review.

1. To run the analysis, click the **Run** icon . Or, right-click the **Curved Beam Analysis - YOUR NAME (-Default-)** folder highlighted in Fig. 17. From the pull-down menu, select **Run**. Either action initiates the solution process.

After a successful solution, a **Results** folder appears at the bottom of the COSMOSWorks manager. This folder should include three sub-folders that contain default plots of results saved at the conclusion of each Study. These folders are named as illustrated in Fig. 17. If these folders do *not* appear, follow steps (a) through (e) outlined on page 1-13 of Chapter 1.

When the **Results** folders are displayed, alter **Units** for the **Stress** and **Displacement** plots, if necessary, as outlined below.

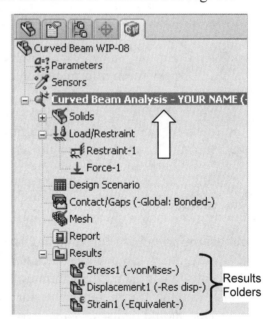

2. Right-click the **Stress1 (-vonMises-)** folder and from the pull-down menu, select **Show**; a stress plot appears.

3. Again, right-click **Stress1 (-vonMises-)** and from the pull-down menu select **Edit Definition…** In the **Display** dialogue box, verify **Units** are set to **psi**. If not change the **Units** from N/m^2 to **psi** using the pull-down menu.

Figure 17 – **Results** folders created during the Solution process.

4. Click **[OK]** ✓ to close the property manager.

5. Repeat steps (2) through (4), however, in steps (2) and (3), right-click on **Displacement1 (-Res disp-)** in place of **Stress1 (-vonMises-)** and in step (3) alter the **Units** field from **m** to **in** (inches).

2-14

Examination of Results

Analysis of von Mises Stresses Within the Model

Outcomes of the current analysis, in the form of plots can be viewed by accessing results stored in the various folders listed in the previous section. This is where validity of results is verified by cross-checking Finite Element Analysis (FEA) results against results based on classical stress equations. *Checking results is a necessary step in good engineering practice!*

1. In the COSMOSWorks manager tree, double-click the **Stress1 (-vonMises-)** folder (or) right-click it and from the pull-down menu, select **Show**. A plot of the vonMises stress distribution throughout the curved beam model is displayed.

Figure 18 reveals an image *similar* to what currently appears on the screen. The following steps convert your current screen image to that shown in Fig. 18.

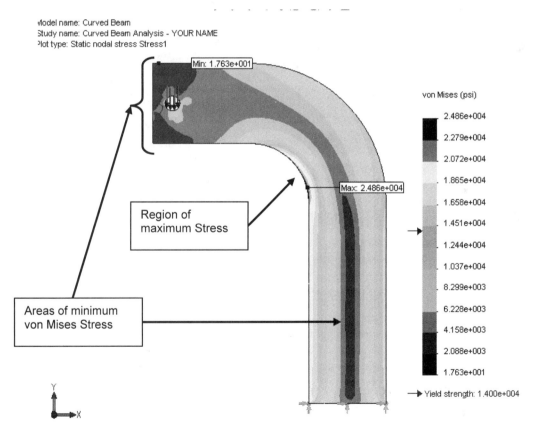

Figure 18 – Front view of the curved beam model showing von Mises stress *after* making changes outlined below. Note arrows indicating Yield Strength on the stress scale at right.

NOTE: *Stress contour plots are printed in black, white, and gray tones. Therefore, light and dark color areas on your screen may differ from images shown throughout this text.*

2. Right-click **Stress1 (-vonMises-)** and from the pull-down menu select **Chart Options….** A portion of the **Chart Options** property manager is shown in Fig. 19.

3. Within the **Display Options** dialogue box, click to place check marks to activate ☑ **Show min annotation** and ☑ **Show max annotation**. This action causes the minimum and maximum von Mises stress locations to be labeled on the model.

4. Click **[OK]** ✓ to close the **Chart Options** property manager.

Figure 19 – Upper portion of **Chart Options** property manager showing current selections.

5. Right-click **Stress1 (-vonMises-)** and from the pull-down menu select **Settings….** The **Settings** property manager opens as illustrated in Fig. 20.

6. From the **Fringe options** menu, select **Discrete** as the fringe type to be displayed.

7. Next, in the **Boundary options** pull-down menu, select **Model** to superimpose an outline of the model on the image.

8. Click **[OK]** ✓ to close the **Settings** property manager.

Figure 20 – Selections in the **Settings** property manager.

9. In the COSMOSWorks toolbar, select the **Stress…** icon. Alternatively, right-click **Stress1 (-vonMises-)** and from the pull-down menu, select **Edit Definition….** Either action opens the **Stress Plot** property manager, shown in Fig. 21.

10. In the **Deformed Shape** property manager, click to "clear" the check-box adjacent to ☐ **Deformed Shape**.

11. Click **[OK]** ✓ to close the **Stress Plot** property manager.

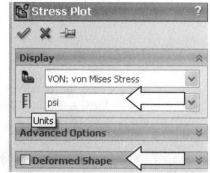

Figure 21 – **Stress Plot** property manager.

The following observations can be made about Fig. 18.

> **OBSERVATIONS:**
> - Areas of low stress (dark blue) occur at the top-left side of the model and also run through the vertical center of the model. The lowest stress is approximately 1.763e+001 = 17.63 psi. Regions of high stress are indicated by red areas. The maximum stress indicated is 2.486e+004 = 24,860 psi, which occurs along the concave surface.
>
> - Material *Yield Strength* ≈ 14000 psi is also listed beneath the color-coded **von Mises** stress color bar. An arrow adjacent to the color bar indicates where the Yield Strength lies relative to all stresses within the model. In this instance, it is clear that some stresses in the model *exceed* the material yield strength. *Yield Strength* and *Safety Factor* are investigated below.

Modern finite element software makes conducting a Finite Element Analysis (FEA) and obtaining results deceptively easy. As noted earlier, however, it is the validity of results and understanding how to interpret and evaluate them that is of primary importance. For these reasons, pause to consider two questions that should be intriguing, or, perhaps, even bothering the reader.

First, why are all stress values positive in Fig. 18 ("+" values typically indicate tension) when compressive stresses are known to exist along the concave surface for the given loading? Second, why does the solution show stresses that exceed the material yield strength when, clearly, stresses above the yield strength indicate yielding or failure? These, and many others, are the types of questions that should be raised continually by users of finite element software. Attempts to address these questions are included below.

To answer these questions, we briefly digress to investigate the definition of von Mises stress and a means to determine a *Safety Factor* predicted by the software.

Von Mises Stress –
The example in Chapter 1 skirted the issue about what the von Mises stress is or what it represents. That example further assumed that some readers might not be familiar with von Mises stress. For the sake of completeness and because von Mises stress typically is not introduced until later in a design of machine elements course, its basic definition is included below. Although this COSMOSWorks user guide is not intended to develop the complete theory related to von Mises stress, the usefulness of this stress might be summed up by the following statement:

> The equation for von Mises stress "allows the most complicated stress situation to be represented by a single quantity."[2] In other words, for the most complex state of stress that one can imagine

---

[2] Budynas, R.G., Nisbett, J. K., Shigley's Mechanical Engineering Design, 8th Ed., McGraw-Hill, 2008, p.216.

(e.g., a three-dimensional stress element subject to a combination of shear and normal stresses acting on every face) these stresses can be reduced to a single number. That number is named the von Mises stress. The number represents a stress magnitude, "which can be compared against the yield strength of the material"[3] to determine whether or not failure by yielding is predicted. As such, the von Mises stress is associated with one of the theories of failure for *ductile* materials; theories of failure are briefly discussed below. The von Mises stress is always a *positive*, *scalar* number.

The above statement answers the question about the positive nature of von Mises stress shown in Fig. 18. It also should provide some understanding about why the von Mises stress can be used to determine whether or not a part is likely to fail by yielding. *Safety Factor* and failure by yielding are explored further below.

Although the above definition indicates that von Mises stress is always a positive number, that superficial answer might continue to bother readers who intuitively recognize that compressive stresses result along the concave surface of the curved beam.

More fundamentally the issue in question gets to the heart of any analysis, whether it be a finite element analysis or stress calculations based on use of classical equations. The answer, of course, is that one must examine the *appropriate stresses* that correspond to the goals of an analysis. For example, in Chapter 1 it was decided that normal stress in the Y-direction (σ_y) was the only stress component that would yield favorable comparisons with stresses calculated using classic equations. It is left as an exercise to determine what the *appropriate stress* is to show variation from compression on the concave surface to tension on the convex side of the curved beam model.

Verification of Results

In keeping with the philosophy that it is *always* necessary to verify the validity of Finite Element Analysis (FEA) results, a quick comparison of FEA results with those calculated using classical stress equations for a curved beam is included below.

Results Predicted by Classical Stress Equations

Although not all users may be familiar with the equations for stress in a curved beam, the analysis below should provide sufficient detail to enable reasonable understanding of this state of stress. The first observation is a somewhat unique characteristic of curved beams, namely that its neutral axis lies closer to the center of curvature than does its centroidal axis. This can be observed in Fig. 22. By definition the centroidal axis,

---

[3] Ibid

Curved Beam Analysis

identified as r_c is located half-way between the inside and outside radii of curvature. However, the neutral axis, identified by r_n, lies closer to the inside (concave) surface. Based upon the above observation, a free body diagram of the upper portion of the curved beam is shown in Fig. 22. Included on this figure are important curved beam dimensions used in the following calculations. Dimensions shown are defined below.

w = width of beam cross-section = 4.00 in (see Fig. 1)
d = depth of beam cross-section = 0.75 in (see Fig. 1)
A = cross-sectional area of beam = $w*d$ = (0.75 in)(4.00 in) = 3.00 in^2
r_i = radius to inside (concave surface) = 3.00 in
r_o = radius to outside (convex surface) = 7.00 in
r_c = radius to centroid of beam = $r_i + w/2$ = 3.00 + 2.00 = 5.00 in
r_n = radius to the neutral axis = $w/\ln(r_o/r_i)$ = 4.00/ln(7.00/3.00) = 4.72 in.
 [determined by equation for a rectangular cross-section of a curved beam]
c_i = distance from the neutral axis to the inside surface = $r_n - r_i$ = 4.72 – 3.00 = 1.72 in
c_o = distance from the neutral axis to the outside surface = $r_o - r_n$ = 7.00 – 4.72 = 2.28 in
e = distance between the centroidal axis and neutral axis = $r_c - r_n$ = 5.00 - 4.72 = 0.28 in

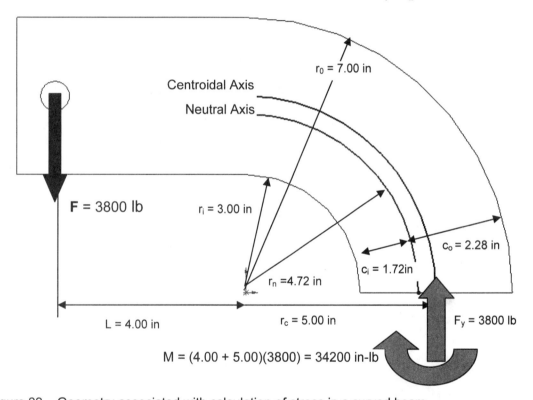

Figure 22 – Geometry associated with calculation of stress in a curved beam.

The reaction force F_y and moment M acting on the cut section are necessary to maintain equilibrium of the upper portion of the curved beam. Equations used to compute the combined bending and axial stresses that result from these reactions are included below. Each equation is of the general form,

$$\text{Curved beam stress} = \pm \text{ bending stress} \pm \text{ axial stress}$$

Where the "$\pm$" sign for bending stress depends on what side of the model is being investigated. Bending stress, caused by moment **M**, is compressive on the concave surface of the curved beam. Hence a minus "-" sign is assigned to the bending stress term in equation [1]. However, on the convex side of the beam bending stress causes tension on the beam surface thereby accounting for the "+" sign associated with the first term in equation [2]. Reaction force **F$_y$** acts to produce a compressive stress on the cut section. Therefore, a minus "-" sign is used with the axial stress component in both equations [1] and [2] below.

Stress at the inside surface:

$$\sigma_i = \frac{Mc_i}{Aer_i} - \frac{F_y}{A} = \frac{-(34200 \text{ in-lb})(1.72 \text{ in})}{(3.00 \text{ in}^2)(0.28 \text{ in})(3.00 \text{ in})} - \frac{3800 \text{ lb}}{3.00 \text{ in}^2} = -24610 \text{ psi} \quad [1]$$

Stress at the outside surface:

$$\sigma_o = \frac{Mc_o}{Aer_o} - \frac{F_y}{A} = \frac{(34200 \text{ in-lb})(2.28 \text{ in})}{(3.00 \text{ in}^2)(0.28 \text{ in})(7.00 \text{ in})} - \frac{3800 \text{ lb}}{3.00 \text{ in}^2} = 11990 \text{ psi} \quad [2]$$

Comparison with Finite Element Results

In addition to serving as a quick check of results, this section reviews use of the **Probe** tool. Both the bending and axial stresses act normal to the cut surface in Fig. 22. Therefore, it is logical that the Finite Element Analysis stress in the Y-direction (σ_y) should be compared with values computed using equations [1] and [2] above. You are encouraged to produce a plot of stress σ_y on your own. However, abbreviated steps are outlined below if guidance is desired.

1. In the COSMOSWorks toolbar, click the **Stress…** icon to open the **Stress Plot** property manager. Alternatively, right right-click the **Results** folder and from the pull-down menu select **Define Stress Plot…**.

2. In the **Display** dialogue box, select **SY: Y Normal Stress** from the pull-down menu. Also in this dialogue box, set the **Units** filed to **psi**.

3. Click to un-check ☐ **Deformed Shape**.

4. Click **[OK]** ✓ to close the **Stress Plot** property manager. A new plot named **Stress2 (-Y normal-)** now appears beneath the **Results** folder and a plot of stress σ_y is displayed. If the plot does not appear, right-click **Stress2 (-Y normal-)** and select **Show** from the pull-down menu.

Next, alter the display to show the mesh and discrete fringes on the model. Proceed on your own and skip to step 8 or, if guidance is desired, follow the steps 5 thorough 7 below.

Curved Beam Analysis

5. Click the **Settings…** icon to open the **Settings** property manager. Alternatively, right-click **Stress2 (-Y normal-)** and from the pull-down menu select **Settings…**.

6. In the **Fringe Options** pull-down menu, select **Discrete** as the fringe display mode and in the **Boundary Options** dialogue box, select **Mesh**.

7. Click **[OK]** ✓ to close the **Settings** property manager.

8. Zoom in on the model to where the curved beam section is tangent to the straight, vertical section as shown in Fig. 23.

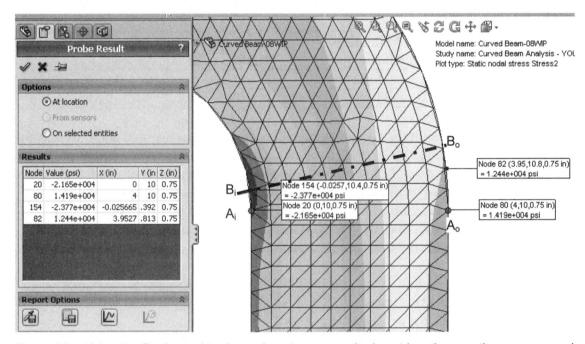

Figure 23 – Using the **Probe** tool to determine stress magnitudes at locations on the concave and convex sides of the curved beam model.

9. In the COSMOSWorks toolbar click the **Probe** tool icon or right-click **Stress2 (-Y normal-)** and from the pull-down menu, select **Probe**. The **Probe Result** property manager opens.

10. In the **Options** dialogue box, select ⊙ **At location** (if not already selected).

11. Move the cursor over the straight vertical edges on the left and right sides of the model. Each edge is highlighted (by a red line) as the cursor passes over it. Click to select two nodes (one on the left and one on the right) located at the *top* of each line. These nodes are located at the *intersection* between the straight vertical section and the beginning of the curved beam section. Selected nodes correspond to A_i and A_o in Fig. 23. If an incorrect node is selected, simply click ⊙ **At**

location in the **Options** dialogue box to clear the selection and repeat the procedure. *Do not close the **Probe Result** property manager at this time.*

The above action records the following data in the **Results** dialogue box: **Node** number, **Value (psi)** of the plotted stress (σ_y), and the **X, Y, Z** coordinates of the selected node. A small "flag" appears adjacent to each node on the model and repeats data listed in the **Results** table. *It may be necessary to click-and-drag column headings to view values in the **Results** table.*

Table I contains a comparison of results found by using classical equations and Finite Element Analysis at these locations.

Table I – Comparison of stress (σ_y) from classical and finite element methods at Section A-A.

Location	Manual Calculation (psi)	Probe Tool Results (psi)	Percent Difference (%)
Point A_i	-24610	-21650	13.7%
Point A_o	11990	14190	15.5%

As an engineer, you should be disappointed and, in fact, quite concerned at the significant difference between the results of these two methods. After all, how can one have confidence in a design when two seemingly rational approaches yield results that differ by 15% or more? When results differ by this order of magnitude it is appropriate to *investigate further* to determine the cause for the disparity and not simply "write off" the differences as due the fact that two alternative approaches are used. Can you provide valid reasons *why* such large differences exist?

Further consideration should reveal the fairly obvious conclusion that St. Venant's principle is once again affecting results. In this instance, a traditional engineering approach would dictate using classical equations for a straight beam in the straight vertical segment of the model below Section A-A, see Fig. 1 (repeated), and a different set of equations in the curved beam portion of the model above Section A-A. However, common sense suggests that there is a transition region between these two segments where neither set of classical equations is entirely adequate. In fact, due to the finite size of elements in this region, it is logical to presume that a Finite Element Analysis provides a more accurate solution in this transition region.

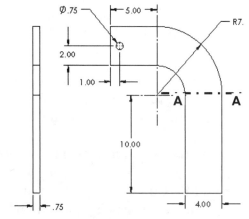

Figure 1 (Repeated) – Basic geometry of the curved beam model.

Given the above observations, we next proceed to sample stress magnitudes at Section B_i-B_o in Fig. 23, which is located slightly above the transition region. Proceed as follows.

12. Move the cursor over the curved edges of the model and on the concave side, click to select the *first* node above the previously selected node.

13. Next, on the convex side of the model, select the *second* node above the previously selected node.

Observe stress magnitudes listed in the **Results** dialogue box and compare them to values listed in Table II below. Nodes **B_i** and **B_o** thus selected lie on a radial line that forms an approximate angle of 7.5° above the horizontal. Stress values calculated using the classical equations are modified to account for a slight shift of the centroidal axis due to beam curvature and for the change in angle of the axial force. Based on these values, a comparison of classical and FEA results in Table II reveals that values differ by at most 4.0%; a significant improvement over the initial calculations.

Table II – Comparison of stress (σ_y) for classical and finite element methods at Section B_i-B_o.

Location	Manual Calculation (psi)	Probe Tool Results (psi)	Percent Difference (%)
Point **B_i**	-24515	-23770	3.1%
Point **B_o**	11940	12440	4.0%

14. Click **[OK]** ✓ to close the **Probe Result** property manager.

This concludes the verification of Finite Element results, but note that even better results would be expected at locations further from the transition region.

Assessing Safety Factor for the Curved Beam

COSMOSWorks provides a convenient means for the stress analyst to determine and view a plot of Factor of Safety distribution throughout the curved beam model. To use this capability, proceed as follows.

1. In the COSMOSWorks toolbar, click the **Design Check Wizard** icon. Alternatively, right-click the **Results** folder and from the pull-down menu, select **Define Design Check Plot…**. The **Design Check** property manager opens as shown in Fig. 24 and displays the first step of a three step procedure.

Figure 24 – **Design Check**, Step #1 of safety factor check.

2. In the **Message** dialogue box read the guidelines for selecting the proper failure criterion to be applied to the current model. Additional details about each failure criterion are provided below.

3. In the upper pull-down menu of the **Step 1 of 3** dialogue box, select **Curved Beam-Extrude1** as the model to be analyzed.

4. Next, in the **Criterion** field, second field from the top, click the pull-down menu to reveal the four failure criteria available for determination of the safety factor.

A brief overview of the four failure criteria is provided below.
- **Max von Mises Stress** – This failure criterion is used for ductile materials (aluminum, steel, brass, bronze, etc.). It is considered the best predictor of actual failure in ductile materials and, as such, provides a good indication of the true safety factor. This criterion is also referred to as the "Distortion Energy Theory" in many design of machine elements textbooks.

- **Max Shear Stress (Tresca)** – This criterion also applies to ductile materials. However, it is a more conservative theory thereby resulting in lower predicted safety factors. As a consequence of its conservative nature, parts designed using this criterion may be somewhat oversized.

- **Mohr-Coulomb Stress** – This failure criterion is applied to the design and analysis of parts made of brittle material (cast iron, concrete, etc.) where the ultimate compressive strength exceeds the ultimate tensile strength ($S_{uc} > S_{ut}$).

- **Max Normal Stress** – Also applicable for brittle materials, this failure criterion does not account for differences between tensile and compressive strengths within COSMOSWorks. This theory is also regarded as the least accurate of the methods available.

5. Because the curved beam is made of aluminum and because a good estimate of safety factor is desired, choose **Max vonMises Stress** (if not already selected) from the pull-down menu.

Also, immediately below this field notice that the Design Check goal is currently defined as

$$\frac{\sigma_{vonMises}}{\sigma_{Limit}} < 1$$

Figure 24 *(repeated)* – First **Design Check** property manager. The comparison stress (failure criteria) is selected here.

In other words, the previous equation is currently set to identify locations in the model where the ratio of von Mises stress to the "limiting" value of stress (i.e., the Yield Strength) is < 1.

Thus, the above criterion identifies locations where yielding of the model is *not* predicted because model Yield Strength, the denominator, is greater than the von Mises stress, the numerator. As initially defined, the above ratio is the *inverse* of the traditional safety factor definition, where:

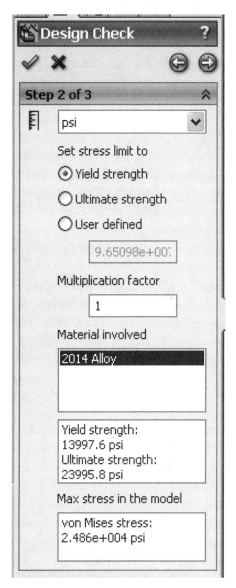

Safety Factor = n = strength/stress

To plot only critical regions of the part, i.e., regions where the Yield Strength is exceeded and the safety factor is < 1, proceed as follows –

6. Proceed to the second step by clicking the right facing arrow button ➡ at the top of this property manager. The **Design Check, Step 2 of 3** dialogue box appears as shown in Fig. 25.

7. In the top pull-down menu, select **psi** as the set of **Units** to be used.

8. Under **Set stress limit to**, click to select ⦿ **Yield strength** (if not already selected).

Notice that the Yield and Ultimate strengths for the model appear near the bottom of this dialogue box. Also, the bottom box lists the maximum von Mises stress found during the solution. By examining the values in these two boxes it is obvious that, somewhere in the part, the maximum von Mises stress *exceeds* the material Yield Strength.

Figure 25 – Step 2 of 3 in the **Design Check** process.

Design Insight

In the event that a *brittle* material is being analyzed by the Mohr-Coulomb or the Max Normal Stress failure criterion, then it is appropriate to select the **Ultimate strength** as the failure criterion.

The **User defined** option is provided for cases where a user specified material is not found in the **Material Property** table or library.

9. Click the right facing arrow button ➡ at top of this property manager to proceed to **Step 3 of 3** in the **Design Check** property manager shown in Fig. 26.

Three options are available for displaying the factor of safety. Brief descriptions of each are provided below.

- **Factor of Safety Distribution** – Produces a plot of safety factor variation throughout the part.

- **Non-dimensional stress distribution** – Compares stress level at any location in the part to the maximum stress level and plots the ratio of these stresses as a number between 0 and 1 throughout the part.

- **Areas below factor of safety** – A target value of safety factor is entered in the field beneath this option. The resulting display shows all areas below the specified safety factor in red and areas with a safety factor greater than specified in blue. This approach easily identifies areas that need to be improved during the design process.

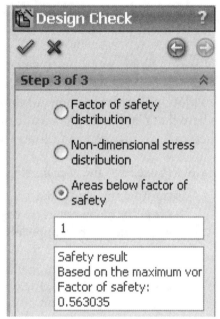

Figure 26 – Redefinition of *Safety Factor* and values to be displayed on the new plot.

10. Beneath **Step 3 of 3**, select ⦿**Areas below factor of safety** and type "1" in the field (if not already "1").

At the bottom of this dialog box the **Safety result** field informs the user that the factor of safety is **0.563035** indicating that the design is *not* safe in some regions of the model. Recall that this value is based on a comparison between Yield Strength and the maximum von Mises stress. (*Values may vary slightly from those shown.*)

Also note that this value of safety factor closely matches that computed by the reciprocal of the equation appearing in the first **Design Check** window. That is:

$$\frac{\sigma_{Limit}}{\sigma_{vonMises}} = \frac{13997}{24860} = 0.56303$$

11. Click **[OK]** ✓ to close the **Design Check** property manager. A new plot folder, named, **Design Check1 (-Criterion: Max von Mises Stress-)**, is listed beneath

the **Results** folder. Also, a plot showing regions of the model where the Safety Factor is < 1.0 (red) and where the Safety Factor is > 1.0 (blue) is displayed.

12. Right-click **Design Check1 (-Criterion: Max von Mises Stress-)**, and from the pull-down menu, select **Chart Options…**. The **Chart Options** property manager opens.

13. In the **Display Options** dialogue box, check ☑ **Show min annotation** and click **[OK]** ✓ to close the **Chart Options** property manager.

The preceding step labels the location of minimum Safety Factor on the curved beam as shown in Fig. 27. As expected, this location corresponds to the location of highest stress previously illustrated in Fig. 18.

The figure now on the screen should correspond to Fig. 27. This figure shows regions where the factor of safety is less than 1 (unsafe regions) in red. Regions with a factor of safety greater than 1 (safe regions) are shown in blue. Localized regions, along the right and left vertical edges and extending into the concave region, have a safety factor less than one.

The line of text, circled near the top-left in Fig. 27, provides a "key" to interpret safe and unsafe regions on the model.

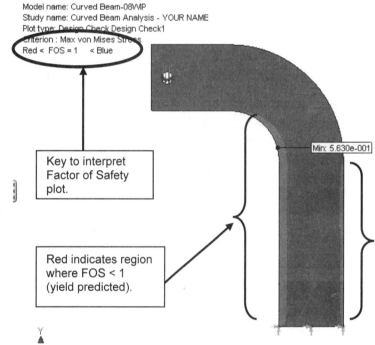

Figure 27 – Curved beam model showing areas where FOS > 1 (safety predicted) and where FOS < 1 (yield predicted).

14. Double-click **Design Check1 (-Criterion: Max von Mises Stress-)** and repeat steps 3 through 11 above, but this time set the **Areas below factor of safety** to **2** instead of **1**, in step 10. How does the plot change?

A designer can repeat this procedure for any desired level of safety factor check.

In summary, an important aspect of the von Mises stress is that it can be used to predict whether or not a part might fail based on a comparison of its stress *value* to the magnitude of yield strength. This topic is aligned with the study of theories of failure.

Analysis of Machine Elements using COSMOSWorks

> **Analysis Insight**
>
> Faced with the above information, a designer would be challenged to redesign the part in any of several ways, depending upon design constraints. For example, it might be possible to change part dimensions to reduce stress magnitudes in the part. Alternatively, if part geometry cannot be changed, a stronger material might be selected or some combination of these or other possible remedies might be applied. Because this aspect of the current example is open-ended, it is not pursued here.

Reaction Forces

It is always good engineering practice to verify that results obtained correlate well with the given information. One simple way to confirm that results "make sense" is to check whether or not reaction forces are consistent with external loads applied to the finite element model. This section examines how to determine reaction forces at the base of the curved beam model. To accomplish this, proceed as follows.

1. In the COSMOSWorks Manager, right-click the **Results** folder and from the pull-down menu, select **List Result Force…**. The **Result Force** property manager opens as shown in Fig. 28.

2. In the **Options** dialogue box, verify that ⊙ **Reaction Force** is selected.

3. In the **Selection** dialogue box, set **Units** to **English (IPS)**, if not already selected.

4. The **Select Faces, Edges or Vertices** field is active (highlighted) and awaiting selection of the entity on which reaction forces are to be determined. Rotate the model so that its bottom (restrained) *surface* is visible and click to select it. **Face<1>** appears in the active field. This is the only face where reactions occur.

5. Click the **[Update]** button and the **Reaction Force (lb)** table at the bottom of the property manager is populated with data.

Figure 28 – Data appearing in the **Result Force** property manager.

The **Component** column of this table lists names for the sum of reaction forces in the X, Y, and Z directions and the Resultant reaction. Force reactions in the X, Y, Z directions found in the **Selection** column are identical to those for the **Entire Model**. This result is expected since the entire model is restrained at only this one location.

Results interpretation is as follows:
    SumX: -0.109   (essentially zero) No force applied to model in X-direction.
    SumY: 3800.2   (essentially 3800 lb = the applied force)
    SumZ: -0.10173   (essentially zero) No force applied to model in Z-direction.
    Resultant: 3800.2   (essentially 3800 lb = the applied force)

It should be noted that a *moment* reaction at the base of the curved beam is missing from the **Reaction Force** table. Also, checking the **Reaction Moment (lb-in)** dialogue box, at the bottom of the property manager, reveals no data entries. This outcome does not agree with the usual conventions for reactions associated with a free-body diagram, but it is consistent with our understanding of **Immovable** restraints applied to three-dimensional, solid, tetrahedral elements. The **Immovable** restraint only restricts translations in the X, Y, Z directions at each restrained node. This observation accounts for the fact that there are only three force reactions and no moments in the **Reaction Force** table of Fig. 28.

6. Click **[OK]** ✓ to close the **Result Force** property manager.

The results above are valid for the **Entire Model**. However, in many instances a model is supported (i.e., restrained) at more than one location. In those instances it is necessary to determine reaction forces at other locations on a model. Performing a reaction check is quite simple and can be viewed as an additional means for users to verify the validity of boundary conditions applied to a model.

Although a surface was selected to examine reaction forces in the above example, it should be evident that other geometric features, such as edges or vertices can also be selected at other locations that might be restrained on a particular model.

Logging Out of the Current Analysis
This concludes an introduction to analysis of the curved beam model. It is suggested that this file not be saved. Proceed as follows.

1. On the Main menu, click **File** followed by choosing **Close** from the pull-down menu.

2. The **SolidWorks** window, Fig. 29, opens and asks, "**Save changes to Curved Beam Analysis?**" Select the **[No]** button. This closes the file without saving any changes.

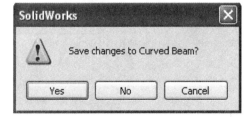

Figure 29 – **SolidWorks** window prompts users to either save changes or not.

EXERCISES

End of chapter exercises are intended to provide additional practice using principles introduced in the current chapter plus capabilities mastered in preceding chapters. Most exercises include multiple parts. Maximum benefit is realized by working all parts. However, in an academic setting, it is likely that parts of problems will be assigned or modified to suit specific course goals.

1. C-clamps, like that illustrated below, must pass minimum strength requirements before they are qualified for general purpose use. Clamps are tested by applying equal and opposite loads acting on the gripping faces. Part of the federal test criteria requires that the movable (lower) jaw be extended a certain percentage of the distance of the fully-open state to ensure that column failure of the screw is an integral part of the test. Presuming that the movable jaw of the clamp satisfies the prescribed test criterion, perform a finite element analysis of the C-clamp subject to the following guidelines.

   Open file: **C-Clamp 2-1**

   - Material: **Cast Carbon Steel** (use S.I. units)

   - Mesh: **High Quality** tetrahedral elements

   - Restraint: **Immovable** applied to the upper gripping surface.

   - Force: **950 N** applied normal to lower gripping surface.

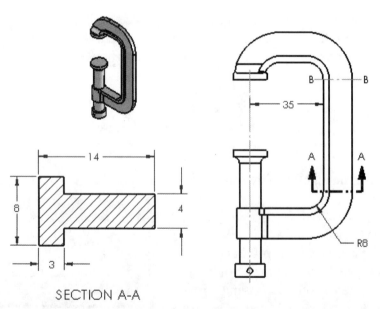

SECTION A-A

Figure E 2-1 – C-clamp frame and cross-section dimensions. Stress to be determined at Section A-A.

Determine the following:
   a. Use classical equations to compute stress at the inside and outside surfaces of the C-clamp frame at section A-A. Section A-A is located where the straight and curved sections are tangent.

   b. Create a stress contour plot of von Mises stress in the frame of the C-Clamp. Include automatic labeling of maximum von Mises stress on this plot.

   c. Use the **Probe** feature to produce a graph of the *most appropriate stress* across section A-A. In other words, because values from this plot are to be compared with manual calculations of part (a), it is necessary to choose the corresponding stress from those available within the finite element software. When using the **Probe** feature, begin at the inside (left) surface and select nodes across the model continuing to the outside of the "T" cross-section. Use equation [1] to compare percent differences between classical and FEA determination of stresses at the inside and outside surfaces.

$$\% \text{ difference} = \frac{(\text{FEA result - classical result})}{\text{FEA result}} * 100 = \qquad [1]$$

   d. Assuming the C-clamp is made of a ductile material, produce a plot showing regions where safety factor < 1.5.

   e. Question: Is there justification for using **High** quality elements for analysis of the C-clamp? Justify your answer by providing reasons either "for" or "against" use of **High** quality elements for this model.

2. A common, metal-cutting "hacksaw" is shown in Fig. E 2-2. A solid model of the hacksaw is available as file: **Hacksaw 2-2**. The model is simplified to include two round holes that pass through the lower-left and right ends of the hacksaw "backbone" labeled in Fig. E 2-2. For analysis purposes, the inside surface of the left-hand hole is to be considered **Immovable**. Use split lines to create a small "patch" of area on the inside surface of the hole located at the right end of the backbone. On this surface apply a 50 lb force induced by a tensile load in the saw blade that is ordinarily held in place between these two holes. Assume the following.

   - Material: **AISI 1020 Steel** (use English units)
   - Mesh: **High quality** tetrahedral elements; use default mesh size.
   - Units: **English (IPS)**
   - Restraint: **Immovable** applied to inside of left hole.
   - Force: **50 lb** applied to inner surface of right-hand hole (split lines needed).

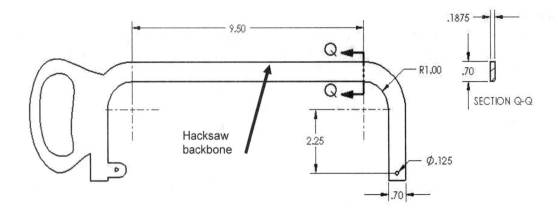

Figure E 2-2 – Basic geometry of a hacksaw frame. Stress is to be determined at Section Q-Q.

Determine the following:
  a. Use classical equations to compute stress at the inside and outside surfaces of the hacksaw frame at section Q-Q. Section Q-Q is located where the straight and curved sections are tangent.

  b. Include a zoomed-in image of the right-hand hole so that the applied load can clearly be seen to act between user specified *Split Lines*.

  c. Create a stress contour plot of von Mises stress in the saw backbone. Include automatic labeling of maximum von Mises stress on the plot.

  d. Use the **Probe** feature to produce a graph of the *most appropriate stress* across section Q-Q, beginning at the inside (concave) surface and continuing to the outside (convex) surface of the backbone cross-section. Use the **Stress Plot** property manager to select the *appropriate stress* for this plot to enable comparison with manual calculations of part (a) above. Also, below the graph, cut-and-paste a copy of the **Probe Results** table showing values used in this comparison. Then use equation [1], repeated below, to compute the percent difference between classical and finite element solutions at the inside and outside surfaces of the saw backbone.

$$\% \text{ difference} = \frac{(\text{FEA result - classical result})}{\text{FEA result}} * 100 = \qquad [1]$$

  e. Based on von Mises stress, create a plot showing all regions of the model where Safety Factor < 2.2 and circle these regions on the plot. Include a software applied label indicating the maximum and minimum values of Safety Factor.

  f. Question: Is there justification for using **High** quality elements for analysis of the hacksaw backbone? Justify your answer by providing reasons either "for" or "against" using **High** quality elements for this model.

3. The curved beam shown in Fig. E 2-3 is subject to a horizontal load applied by means of a pin (not shown) that passes through a hole in its upper end. A solid model of this part is available as file: **Anchor Bracket 2-3**. The lower-left end of the part is attached to a rigid portion of a machine frame (not shown). Thus, because three-dimensional tetrahedral elements are to be used to model this part, the restraint at this location should be considered **Immovable**. Use split lines to create a small "patch" of area on the inside surface of the 16 mm diameter hole. Locate these split lines 24 mm from the right edge of the model. On this surface apply a horizontal force of 8600 N acting in the positive X-direction (to the right). Assume the following.

- Material: **AISI 1010 Steel, hot rolled bar** (use SI units)

- Mesh: **High Quality** tetrahedral elements; use default mesh size.

- Restraint: **Immovable** applied on the inclined surface.

- Force: **8600 N** in the X-direction applied on the right, inside surface of the hole.

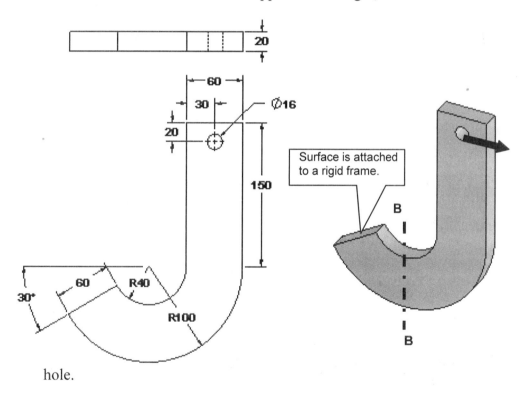

Figure E 2-3 Dimensioned view of the Anchor Bracket. Stress is to be determined at Section B-B.

Determine the following:
  a. Use classical equations to compute stress at the inside (concave) surface and the outside (convex) surface of the anchor bracket at section B-B. Section B-B passes through the center of curvature of the curved beam and is considered to be a vertical line.

b. Include a zoomed-in image of the hole so that the applied load can clearly be seen to act between user specified *Split Lines*.

c. Create a stress contour plot of von Mises stress in the saw backbone. Include automatic labeling of maximum von Mises stress on the plot.

d. Using von Mises stress, create a plot showing all regions of the model where Safety Factor < 1.0. Indicate regions (if any) where the Safety Factor < 1 occurs by circling their location(s) on the figure and labeling them as FOS < 1. Include a software applied label indicating locations of maximum and minimum values of Safety Factor.

e. Use the **Probe** feature to produce a graph of the *most appropriate stress* across section B-B, beginning at the inside (concave) surface and continuing to the outside (convex) surface of the bracket cross-section. Use the **Stress Plot** property manager to select the *appropriate stress* for this plot to enable comparison with manual calculations of part (a) above. Also, below the graph, cut-and-paste a copy of the **Probe Results** table showing values used in this comparison. Use equation [1], repeated below, to compute the percent difference between classical and finite element solutions at the inside and outside surfaces of the bracket.

$$\% \text{ difference} = \frac{(\text{FEA result - classical result})}{\text{FEA result}} * 100 = \quad\quad [1]$$

HINT: Because the mesh generation scheme within COSMOSWorks creates an optimized mesh, it is probable that (a) a straight line of nodes does *not* exist across Section B-B (choose the best straight line), and (b) it is also *unlikely* that node points *occur exactly on a vertical centerline through the center of curvature.* For these reasons, and to obtain the best estimate of stress at on a vertical line through the curved section, proceed as follows.

- Zoom in on the curved section of the model.

- Open the **Probe Result** property manager. Simultaneously the SolidWorks flyout menu appears. Click the "+" sign to display this menu.

- In a front view of the model, move the cursor over the **Right Plane** in the flyout menu. This action highlights an edge view of the **Right Plane** to assist in locating nodes closest to a vertical line at Section B-B.

Textbook Problems
In addition to the above exercises, it is highly recommended that additional curved beam problems be worked from a design of machine elements or mechanics of materials textbook. Textbook problems provide a great way to discover errors made in formulating a finite element analysis because they typically are well defined problems for which the solution is known. Typical textbook problems, if well defined in advance, make an excellent source of solutions for comparison.

CHAPTER #3

STRESS CONCENTRATION ANALYSIS

This example explores stress in the vicinity of a geometric discontinuity in a part where stress concentration is known to occur. Because geometric discontinuities can assume a variety of shapes, they are often generically referred to as "notches." Stress concentration and their related stress concentration factors are typically studied in mechanics of materials and/or design of machine elements courses. Therefore, this example focuses on the validity of Finite Element Analysis (FEA) solutions in the vicinity of these geometric discontinuities and on the effects of mesh size on solution accuracy. However, the general principles learned in this example apply to a wide variety of finite element analysis problems. More specifically, since it is generally accepted that improved finite element solutions result when a smaller mesh size is used, this example examines convergence to a solution through the application of successive mesh refinement.

Because this is the third example, fewer figures and briefer step-by-step procedures are included. In fact, where you already are familiar with procedures and where *no new* procedures are introduced, special instructions encourage users to complete specific tasks on their own. The combination of these two approaches should permit users to work more at their own pace while allowing expanded discussion of new topics.

Learning Objectives
In addition to software capabilities mastered in previous chapters, upon completion of this example, users should be able to:

- Recognize when, how, and why to *defeature* a model.

- Apply *mesh refinement* to a model and understand the influence of mesh density on stress and displacement results.

- Create *multiple copies* of *related studies* quickly and easily within COSMOSWorks.

- Check *convergence* to gauge validity of a FEA solution.

- Use multiple *viewports* to compare results of different FEA analyses.

Problem Statement
The rectangular bar illustrated in Fig. 1 is fixed at its left-end and is subject to an axial, tensile load **F = 6000** lb applied to the opposite end. A rounded "notch," which causes stress concentration, is located in the center of the member.

Figure 1 – Axially loaded bar with a geometric discontinuity.

Dimensions of the notched bar, shown in Fig. 2, reveal that the member is sufficiently long so that boundary-conditions (i.e., restraints and loads) do not significantly affect stresses near the notch. The bar is made of **AISI 1020** steel. Analysis of this model begins below.

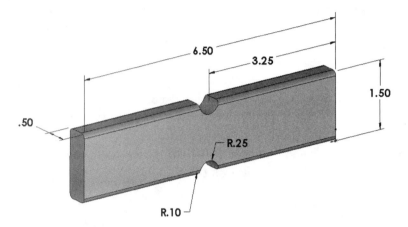

Figure 2 – Dimensioned drawing and a three-dimensional model of a notched bar.

1. Open SolidWorks by making the following selections. (*Note:* "/" is used to separate successive menu selections.)

   Start / All Programs / SolidWorks 2008

2. When SolidWorks is open, select **File / Open**. Then use procedures common to your computer environment to locate and open the COSMOSWorks file named **"Notched Bar."**

Create a Static Analysis (Study)

The Study to be created is to be named "**Draft Quality Mesh DEFAULT.**" Individuals comfortable with setting up a **Static** analysis using a **Solid Mesh** are encouraged to proceed on their own. Abbreviated steps, 1 through 4 below, are provided for those desiring guidance. *NOTE: This and future chapters make use of right-clicking menu items to make selections. Users who prefer to use COSMOSWorks icons, as demonstrated in Chapter 2, are encouraged to do so.*

1. Click the **COSMOSWorks** analysis manager icon to switch from SolidWorks to COSMOSWorks.

2. Right-click **Notched Bar** and from the pull-down menu and select **Study…**. The **Study** property manager opens.

3. In the **Name** dialogue box, change the name **Study 1** by typing "**Draft Quality Mesh DEFAULT**" where the word "DEFAULT" refers to the mesh size used.

4. Verify that a **Solid Mesh** and **Static** study are selected, then click **[OK]** ✓. An outline for the study is opened in the COSMOSWorks manager.

Defeaturing the Model

Before beginning an analysis, it is always a good idea to examine the model to determine whether or not its geometry can be simplified without significantly impacting the analysis. The reason for this is that a simpler model results in a more computationally efficient analysis. Thus, the user must ask, "Is it necessary to include all geometric features on the model in order to obtain a valid solution?" If the answer is "no," then the model can be simplified by *suppressing* (i.e., defeaturing) unimportant features. Typical items that can be suppressed without significantly affecting results are minor geometric features, such as fillets, rounds and chamfers. *You are cautioned, however, that defeaturing a model can have dire consequences if improper choices are made.* An example of a feature that should *not* be removed is examined in a future example. Defeature the current model as described by two alternate methods included below. It is strongly recommended that both methods be attempted.

Method 1:
1. Depress and *hold* the **Shift** key and move the cursor onto the graphics screen. Then click to select each of the eight rounded surfaces on the model. Some of these surfaces are shown shaded in Fig. 3.

2. After selecting all rounds, right-click anywhere in the graphics screen and the pop-up menu shown in Fig. 3 appears. On this menu, select the **Suppress** icon circled in Fig. 3. The suppressed rounds are effectively "removed" from the model as illustrated in Fig. 4

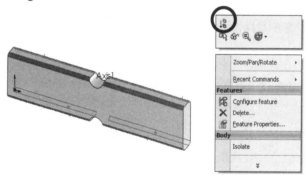

Figure 3 – Rounds selected to be suppressed on the Notched Bar model.

The above procedure works well when only a few geometries are chosen for defeaturing. However, an alternative and somewhat simpler means can be used when the goal is to suppress all rounded edges on this model. Next, use the alternative approach, but first the rounded edges must be returned to the model to enable demonstrating the alternate method. Proceed as follows.

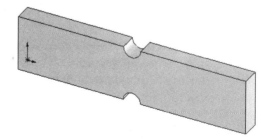

Figure 4 – Notched Bar after defeaturing.

3. Click the **SolidWorks** feature manager icon to return to SolidWorks. NOTE: In future instructions, switching back-and-forth between COSMOSWorks and SolidWorks is indicated by the word "toggle."

4. In the SolidWorks manager tree, locate **Fillet1** and right-click this icon. Above the pop-up menu select the **Unsuppress** icon. Fillets reappear in the model.

Method 2:
5. Once again, locate **Fillet1** in the SolidWorks manager tree and right-click this label. From above the pop-up menu, select the **Suppress** icon. All fillets are removed from the model.

6. Toggle back to COSMOSWorks by selecting its icon.

Analysis Insight:
Clearly **Method 2** is the easier of the two methods. However, it requires planning ahead when the SolidWorks model is created. For example, if rounds of the same size were included elsewhere on a more complex model, but if only one set of rounds were to be suppressed, then the SolidWorks part should be created with two different sets of rounds each of which is easily identified by a different name.

Assign Material Properties to the Model

If possible, specify the material for this model on your own. Choose **AISI 1020** steel and select **English (IPS)** units. An abbreviated, step-by-step procedure is provided below if guidance is needed.

1. Right-click the **Solids** folder and from the pull-down menu select **Apply Material to All…**. The **Material** window opens.

2. Select ⦿ **From library files**, then from the pull-down menu select **cosmos materials** (if not already selected).

3. Next, click the "+" sign adjacent to **Steel (30)** and select **AISI 1020** steel from the list of available steels. *Caution: Avoid selecting AISI 1020 Steel, Cold Rolled.*

4. On the **Properties** tab, adjacent to **Units**, select **English (IPS)**.

5. Click **[OK]** to close the **Material** widow. A check mark "✓" appears on the **Solids** folder.

Stress Concentration

Apply Restraints and Loads

Loads and restraints acting on this model include an **Immovable** restraint applied on its left-end and a **6000** lb axial force normal to its right-end. Practice applying these restraints and loads on your own. After applying restraints and loads, the COSMOSWorks Manager should appear as shown in Fig. 5 and the model should appear as shown in Fig. 6. An abbreviated, step-by-step procedure is provided below if guidance is needed.

1. In the COSMOSWorks manager, right-click **Load/Restraint** and select **Restraints….** The **Restraint** property manager opens.

2. Under **Type**, select **Immovable (No translation)**. The **Faces, Edges, Vertices for Restraint** field is highlighted (light blue) and is awaiting selection of the surface to be designated **Immovable**.

3. Rotate the model as necessary and select the left-end surface of the model. **Immovable** restraints are applied to the model as seen at left in Fig. 6.

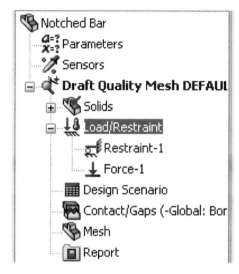

Figure 5 – COSMOSWorks analysis manager with **Load/Restraint** defined.

4. Click **[OK]** ✓ to close the **Restraint** property manager. **Restraint-1** appears beneath the **Load/Restraint** icon as seen in Fig. 5.

5. Once again right-click the **Load/Restraint** icon, but this time select **Force….** The **Force** property manager opens.

6. Beneath **Type**, click to select ⦿ **Apply normal force**.

7. Immediately below the previous selection, the **Faces and Shell Edges for Normal Force** field is highlighted. Rotate the model and select the right-end surface. Force vectors appear on the right-end as shown in Fig. 6.

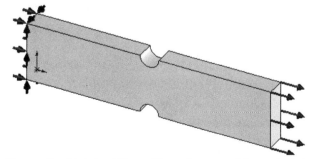

Figure 6 – Notched Bar with an **Immovable** restraint on its left-end and a **6000** lb tensile force applied normal to its right-end.

3-5

8. Verify that the **Units** field is set to **English (IPS)**.

9. In the **Normal Force/Torque (Per entity)** field, type **6000**.

10. Because the default orientation for forces applied normal to a surface is directed *toward* the surface, it is necessary to check ☑ **Reverse direction** to apply a tensile load.

11. Click **[OK]** ✓ to close the **Force** property manager. **Force-1** is now listed beneath the **Load/Restraint** icon in Fig. 5.

The model should now appear as shown in Fig. 6.

Meshing the Model

Because a primary goal of this example is to examine differences between results obtained when a different mesh *type* or mesh *size* is used, it is suggested that the following mesh definition steps be followed carefully. The model will be meshed five times, each time using a different size or type of mesh or mesh control capability. This example concludes with a comparison of results obtained when using different meshes and corresponding observations about solution accuracy.

Proceed as follows to define the first mesh.

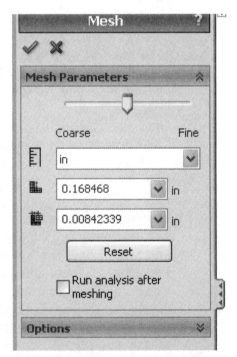

1. Right-click the **Mesh** icon and select **Create Mesh…**. The **Mesh** property manager opens as shown in Fig. 7.

2. In the **Mesh Parameters** dialogue box, verify that **Unit** is set to **in**. Accept remaining default settings for mesh size and tolerance. A pointer at the top of this dialogue box should be located approximately in the center of the scale between **Coarse** and **Fine** mesh sizes. This is the default mesh size.

3. At bottom of the **Mesh** property manager, click ⌄ to open the **Options** dialogue box, which is shown in Fig. 8.

Figure – 7 Default settings for mesh size in the **Mesh** property manager.

4. In the **Options:** dialogue box, change the mesh **Quality** to a ⦿**Draft** quality mesh (unless already selected).

5. Verify other system default settings appear as shown in Fig. 8 and click **[OK]** ✓ to close the **Mesh** property manager.

The meshed model appears in Fig. 9.

Figure 10 shows an enlarged view of the notch and the resulting straight-line approximation of the curved notch that results when a **Draft** quality mesh is used.

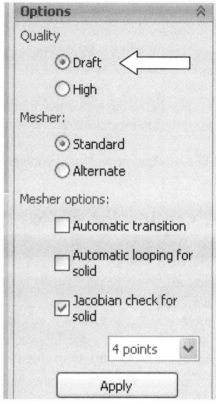

Figure 8 – **Options:** portion of the **Mesh** property manager. Mesh **Quality** is set to **Draft**.

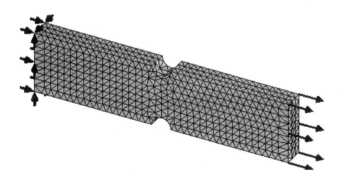

Figure 9 – Draft quality mesh shown on the Notched Bar model.

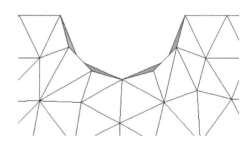

Figure 10 – Close-up view of **Draft** quality mesh reveals straight-line approximation of the curved surface.

Analysis of Machine Elements using COSMOSWorks

Solution

Proceed directly to the solution. You are encouraged to run the **Study** on your own. An abbreviated, step-by-step procedure is provided below in the event guidance is desired.

1. In the COSMOSWorks analysis manager, right-click the study name **Draft Quality Mesh DEFAULT (-Default-)**.

2. From the pull-down menu, select **Run**. The analysis runs and the usual **Results** folders are added at the bottom of the COSMOSWorks Manager.

The solution process is complete. Next, proceed to examine the results of this analysis.

Examination of Results

Stress Plots

Recall that this example will be solved five different times, each time using a different mesh size, mesh type or mesh control technique. Thus, a second goal of this example is to introduce a simple method of creating additional studies that are identical to the first except for different mesh characteristics. For this reason, it is necessary to consider carefully exactly *what results* are to be examined and what characteristics are desired for the *display* of those results. If these two factors are addressed when the *first* set of results is being defined, then similar sets of results and displays can automatically be produced for all subsequent solutions. Proceed as follows to specify the desired plot characteristics used to display the output results.

1. Beneath the **Results** folder in the COSMOSWorks Manager tree, right-click **Stress1 (-vonMises-)** and, from the pull-down menu, select **Show**. This action displays the vonMises stress contour plot and accompanying color-coded stress legend. Alternatively, double-click **Stress1 (-vonMises-)** to display the plot.

The following steps are used to specify certain desired characteristics of the current screen image. While some selections are arbitrary, to demonstrate how solution output display can be controlled from one study to the next, other selections represent the author's preferences. In each case a brief justification is given. In practice, however, many of these selections depend upon user preference and/or standard practices within a specific company or industry. As selections below are made, an image of the Notched Bar gradually begins to appear like that shown in Fig. 11.

2. Right-click **Stress1 (-vonMises-)** and from the pull-down menu, select **Edit Definition...**. The **Stress Plot** property manager opens.

3. In the **Display** dialogue box, set **Units** to **psi**.

3-8

4. At the bottom of this property manager, click ▼ to open the **Property** dialogue box (if not already open).

5. Click to check ☑ **Include title text:** and type **Your Name** and any other information pertinent to this plot. *Justification:* It is important to document the author and context of a Study so that other users of this information know whom to contact if questions arise.

6. Click **[OK]** ✓ to close the **Stress Plot** property manager. The information just entered (your name, etc.) appears at the top-left of the graphics screen along with default information about the plot. Briefly examine information included there.

7. Right-click **Stress1 (-vonMises-)** again. From the pull-down menu, select **Chart Options…**. The **Chart Options** property manager opens.

8. Beneath **Display Options** click to check ☑ **Show max annotation**. The location of maximum stress is immediately labeled on the plot as shown near the top notch in Fig. 11. *Justification:* This choice is made because it is desired to determine the location of maximum stress occurring in the part and whether or not it is near the geometric discontinuity.

9. Click **[OK]** ✓ to close the **Chart Options** property manager.

10. Right-click **Stress1 (-vonMises-)** and from the pull-down menu, select **Settings…** The **Settings** property manager opens.

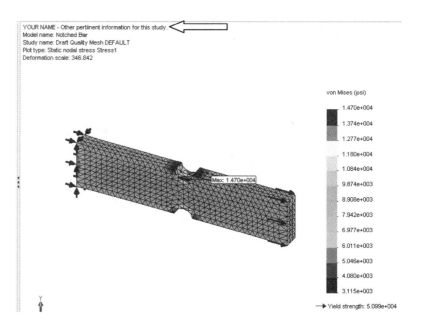

Figure 11 – Trimetric view of the Notched Bar with user specified display options.

11. Beneath **Fringe options**, select **Discrete**. *Justification:* This option is selected because discrete fringes reproduce better when printed in black, white and gray tones used in this text. Otherwise, fringe display is a user preference.

12. Beneath **Boundary options**, select **Mesh**. This option displays the current mesh on the model. *Justification:* Since one goal of this example is to investigate effects of mesh modification upon results, viewing the mesh provides visual feedback about mesh size.

13. Click **[OK]** ✓ to close the **Settings** property manager.

A completed view of the model displaying von Mises stress contours appears in Fig. 11. Note the maximum magnitude of vonMises stress determined by this analysis. **Table 1**, near the end of this example, summarizes this value and other results obtained when different mesh sizes and mesh types are applied to the model. A comparison of results is deferred until later.

Analysis Insight

Two views of the Notched Bar showing discrete vonMises stress contours are displayed in Figs. 12 and 13. For these plots, the following observations are made.

OBSERVATIONS for Fig. 12:
- The magnitude and distribution of stress at left and right ends of the model are different due to the **Immovable** restraint applied at the left end and uniform force applied normal to the right-end. However, due to length of the model, these effects are minimized near the middle of the model (the area of primary interest).

- A comparison of deformed and *un*-deformed shapes of the model can be made by moving the cursor over the model. This action causes an outline of the *un*- deformed shape to be superimposed on the model. In the front view of the model in Fig. 12, a slight narrowing in the Y-direction is observed by noting a small gap between the model outline and the stress contour plot. This change of lateral dimension is due to Poisson's ratio effect.

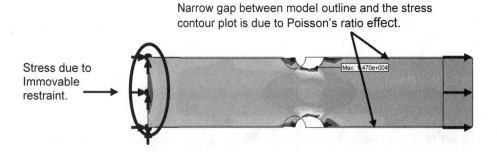

Figure 12 – vonMises stress contours and comparison of deformed and un-deformed models of the Notched Bar. The **Mesh** is not shown on this plot so that Poisson's effect can be seen.

Analysis Insight (continued)

OBSERVATIONS for Fig. 13:
- Superimposing the current mesh onto the model reveals slight differences between element geometry near the top and bottom notches. Carefully observe these differences on Fig. 13 or on your screen.

- Stress *distribution* adjacent to notches located at the top and bottom of the bar is different. This difference may be due to somewhat different mesh shapes in these two areas of the model.

NOTE: Because the meshing process is automated and because it "starts form scratch" and seeks an optimum mesh each time a part is meshed, it is possible that a user obtained mesh might differ from that shown in Fig. 13.

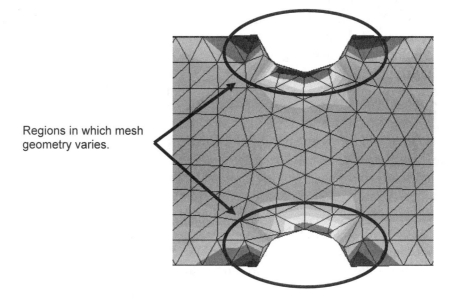

Figure 13 – Close-up view of **Draft** quality mesh in the vicinity of the two notches. Carefully observe slight differences in mesh geometry and distribution of vonMises stress.

Creating a Copy of a Plot

Due to the nature of the axial force applied to the Notched Bar (an axial force in the X-direction), it is also prudent to examine stresses in the X-direction (i.e., σ_x). To do so requires definition of another **Stress Plot** similar to that defined for the vonMises stress plot displayed above. However, repeating all steps required to produce a plot with the *same* characteristics is somewhat tedious and time consuming. Therefore, use the following shortcut to create a *copy* of the current graph and all its display settings.

1. In the COSMOSWorks analysis manager click-and-drag the **Stress1 (-vonMises-)** folder upward and drop it onto the **Results** folder as illustrated in Fig. 14.

   A new copy of the contents of the **Stress1 (-vonMises-)** folder appears at the bottom of the **Results** list and is labeled **Copy [1] Stress1 (-vonMises-)**. This new folder contains an exact duplicate of all commands previously defined for the **Stress1 (-vonMises-)** folder. The following steps outline how to modify this *copied* folder to contain a plot of σ_x.

2. Click-*pause*-click on the name **Copy [1] Stress1 (-vonMises-)**. And type a new descriptive name for the new plot. Since this new folder is to contain σ_x, type: "**Sigma-X**" and press **[Enter]**.

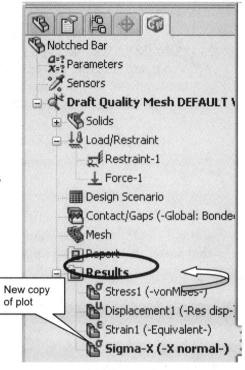

Figure 14 – Making an exact copy of a plot folder.

The *changed name* now appears as **Sigma-X (-vonMises-)**. This name, of course, makes no sense since these names represent two different stresses. The reason for this is that, although the folder *name* was changed, the copied contents (namely vonMises stress) still reside in the folder. However, this is easily corrected by the following steps.

3. Right-click **Sigma-X (-vonMises-)** and the pop-up menu at right displays possible options. Select **Show** and the existing vonMises stress plot is displayed again. Do nothing with this plot.

4. Right-click the **Sigma-X (-vonMises-)** folder a second time. This time the traditional pull-down menu appears. A *partial* view of this menu is shown in Fig. 15. From this menu select **Edit Definition…** and the **Stress Plot** property manager opens.

5. In the dialogue box beneath **Display**, open the **Component** pull-down menu and from the list of possible stresses select **SX: X Normal stress**.

Figure 15 – Partial view of pull-down menu.

6. Click **[OK]** ✓ to close the **Stress Plot** property manager.

The *correct* name for the stress now appears in the COSMOSWorks manager tree as **Sigma-X (-X normal-)**. This name is a combination of the name you typed to identify the stress plus a system applied label to identify σ_x.

7. A plot of **Sigma-X (-X normal-)** should appear on the screen as seen in Fig 16. If not, click **Sigma-X (-X normal-)** and select **Show**.

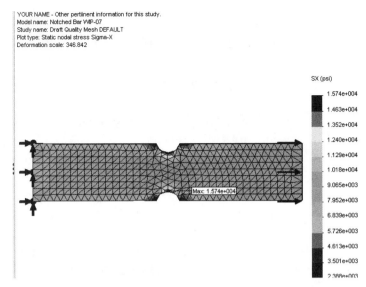

Figure 16 – Graphic display of normal stress in the X-direction (σ_x).

Analysis Insight
OBSERVATIONS for Fig. 16:

- Maximum stress in the X-direction σ_x is larger than the vonMises stress. σ_x = 15,740 psi while maximum vonMises stress is σ' = 14,700 psi. Thus, in this instance it is worthwhile to examine stress in the X-direction.

- All display settings established for **Stress1 (-vonMises-)** are replicated in the **Sigma-X (-X normal-)** plot folder, (i.e., discrete fringes are shown, your name and a brief description appear at top-left of the graphics screen, the maximum stress magnitude and its location are labeled on the plot, and the mesh is superimposed on the model). NOTE: If the description appearing next to your name in the top line is no longer valid, it can be changed; see steps 2 through 5 of the previous section.

Displacement Plot

Another goal for this example is to determine when an analysis converges to a solution. Although both stress and displacement can be used to make this determination, it turns out that displacement plots typically yield better information about solution convergence than do stress contour plots. Based on your study of mechanics of materials, do you understand the reason for this? This question is answered later in this example. Based on this discussion, a **Displacement** plot is created next.

1. Double-click the **Displacement1 (-Res disp-)** folder. The default displacement plot named **Displacement1 (-Res disp-)**, which is short for "resultant displacement," is displayed.

The resultant displacement image, **(-Res disp-)**, is simply a plot of the square root of the sum of the squares of the X, Y, and Z displacement components as given by equation [1] below.

$$\text{Resultant Displacement} = \sqrt{X^2 + Y^2 + Z^2} \qquad [1]$$

It is entirely acceptable to use **(-Res disp-)** for this analysis. However, we next select the X-displacement to be displayed for two reasons. First, the X-displacement is in the direction of the applied load and therefore it accounts for the primary displacement component (Y and Z displacements due to Poisson ratio effects are *very* small relative to X). And second, choosing the X-displacement provides an opportunity to demonstrate selection of a different displacement component as outlined in the following steps.

2. Right-click **Displacement1 (-Res disp-)**, and from the pull-down menu select **Edit Definition…**. The **Displacement Plot** property manager opens as shown in Fig. 17.

3. In the **Display** dialogue box, open the **Component** pull-down menu and select **UX: X Displacement** to display the X-component of displacement.

4. Verify that **Units** are set to inches **in**.

5. Click **[OK]** ✓ to close the **Displacement Plot** property manager.

The X-displacement image is displayed using system default settings. Below, two cosmetic changes are made to this plot.

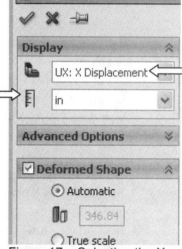

Figure 17 – Selecting the X-component of displacement for display.

6. Right-click **Displacement1 (-X disp-)** and from the pull-down menu, select **Settings…** The **Settings** property manager opens as shown in Fig. 18.

7. Beneath **Fringe options** select **Discrete** (if not already selected) for the type of fringe display and beneath **Boundary options** select **Model** (if not already selected) to turn on a display of the model outline. A mesh display is not needed here.

Figure 18 – Display options specified in the **Settings** property manager.

3-14

Stress Concentration

8. Click **[OK]** ✓ to close the **Settings** property manager and a plot of **UX: X Displacement** is displayed like that shown in Fig. 19.

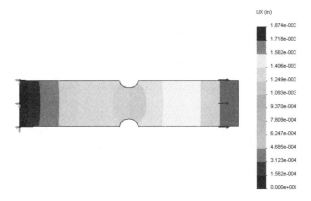

Figure 19 – Displacement plot modified to show **UX: X Displacement** using an outlined model and discrete fringes.

Recall that default plots show exaggerated displacements. However, if it is desired to view *true* displacements, right-click **Displacement1 (-X disp-)** and from the pull-down menu select **Edit Definition…**. The **Displacement Plot** property manager, shown in Fig. 17, opens and beneath **Deformed Shape** click to select ⦿ **True scale** and click the **[OK]** ✓. This action establishes a 1:1 scale for the displacement plot. Try this; then before proceeding, change back to the ⦿ **Automatic** default setting.

9. Click **[OK]** ✓ to close the **Displacement** property manager.

Because both stress and displacement plots are defined above, it is possible to produce additional *nearly* identical *copies* of the entire Study created above. A copy of this Study can be created in which everything remains the same except for mesh type and size. Upon completion of additional studies using different mesh types and sizes, the stress and displacement results from all studies will be compared. Proceed as follows to define these new studies

Creating New Studies

Study Using High Quality Elements and COARSE Mesh Size

The procedure below outlines how to create a new Study based upon specifications defined in an earlier study. To do this, the Study to be copied must currently be open. There are two ways to create a new study; both are described below. It is suggested that the **First Method** be used at this time. However, also read the **Second Method** and consider how it parallels making a copy of the von Mises plot demonstrated above.

First Method:
1. In the COSMOSWorks analysis manager, right-click **Draft Quality Mesh DEFAULT (-Default-)**. From the pull-down menu, select **Copy**.

3-15

2. Move the cursor to the top of the COSMOSWorks manager tree and right-click **Notched Bar**. From the pull-down menu, select **Paste**. The **Define Study Name** window opens as shown in Fig. 20.

> **Second Method:**
> The second way to create a copy of the current Study is to click-and-drag the **Draft Quality Mesh DEFAULT (-Default-)** folder and place it onto **Notched Bar** located at the top of the COSMOSWorks manager tree. This action also opens the **Define Study Name** window shown in Fig. 20.

Either method for making a copy leads to the same steps outlined below.

3. On the **Study Name:** line of this window, type: "**High Quality Mesh- COARSE**." A 30 character limit applies to the length of Study names.

4. Click **[OK]** to close the **Define Study Name** window.

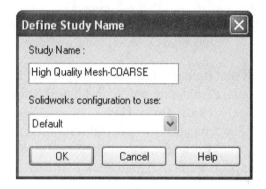

Figure 20 – A new Study name must be typed into the **Define Study Name** window.

A new Study folder named **High Quality Mesh-COARSE (-Default-)** appears immediately beneath the original Study folder in the manager tree.

5. If necessary, click the "+" sign adjacent to this folder and observe that all portions of the new study are pre-defined based upon contents of the original **Draft Quality Mesh DEFAULT (-Default-)** Study.

Because the contents of both studies are currently *identical copies* of one another, except for their Study names, proceed as follows to alter the mesh type to match the name just assigned to this new Study.

6. Beneath **High Quality Mesh – COARSE**, right-click the **Mesh** folder and from the pull-down menu, select **Create Mesh…**. A COSMOSWorks warning window opens, see Fig. 21. It displays the warning "**Remeshing will delete the results for study: High Quality Mesh-COARSE**."

7. Click **[OK]** to delete the current results. The **Mesh** property manager opens.

Recall that existing results contained in the newly created **High Quality Mesh-COARSE** Study are still those *copied* from the previous study when a **Draft** quality mesh was used. Thus, those results are to be replaced by results of the new study as defined below.

Stress Concentration

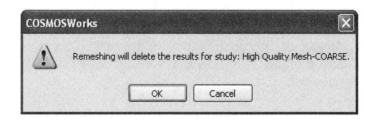

Figure 21 – Warning alerts the user that results associated with the current mesh will be deleted.

8. In the **Mesh** property manager, note the current mesh size (0.168468 in). Then, beneath **Mesh Parameters** click-and-*drag* the mesh size slide-control to the **Coarse** label at the far left-side of the scale. Mesh size is now (0.33693541 in), which is twice as large as the default mesh indicated at the middle of the scale.

9. Next, at the bottom of the **Mesh** property manager, select ⌄ to open the **Options:** dialogue box.

10. Beneath **Quality** click to select ⦿ **High**. All other settings in this window remain unchanged as shown in Fig. 8; these are the default settings.

Scroll back up to the **Mesh Parameters** dialogue box and notice that changing the mesh type from **Draft** quality to **High** quality does *not* change the mesh size. The coarse mesh size is still 0.33693541 in. Using a **High** quality mesh simply adds a mid-side node to each side of the tetrahedral element, thereby making it better able to model curved surfaces. This added ability to model curvature makes **High** order elements more *flexible* (i.e. less stiff) than **Draft** quality elements. The effects of this change are examined later in this example.

11. Click **[OK]** ✓ to close the **Mesh** property manager.

Re-meshing of the model occurs immediately and when complete, the COSMOSWorks manager appears as shown in Fig. 22. Notice the *Error* symbol adjacent to the **High Quality Mesh-COARSE (-Default-)** study name. A *Warning* symbol also appears adjacent to the **Results** folder. Descriptions of the three warning and/or error symbols used in COSMOSWorks are described next.

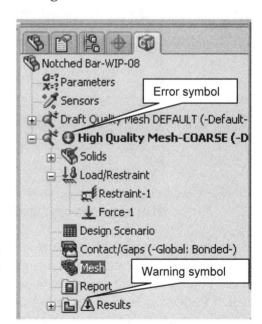

Figure 22 – COSMOSWorks Analysis Manager tree showing an error and several warnings.

3-17

Icon	Description
⊙	This icon indicates an error in the Study. It appears adjacent to the Study name near the top of the COSMOSWorks manager tree and on the feature that contains the error.
⊗	This icon indicates an error with a feature of the study. It appears on the feature name in the COSMOSWorks manager tree.
⚠	This icon indicates a warning concerning a feature. It appears on the specific feature or on the **Results** folders if they are invalid.

12. To determine the source of warnings and errors, right-click the **High Quality Mesh-COARSE (-Default-)** study name and from the pull-down menu, select **What's wrong?...**. The **What's Wrong** window opens as shown in Fig. 23.

Corresponding to the **Results** folder *warning*, a brief description of the error states:

"**Results may not be valid due to the following changes to this study: Mesh information has been changed**."

This message warns that existing results in the copied folders are no longer valid because they correspond to the original Draft quality mesh. However, a new, high-quality mesh was just defined above. Therefore, a new solution must be run using this new mesh.

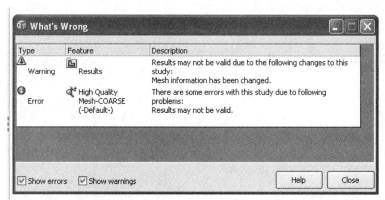

Figure 23 – Illustration of a warning and error listed in the **What's Wrong** error reporting window.

Next, read the *error* message corresponding to the Study name. It reads: "**Results may not be valid**." Like most error messages, this message is not overly revealing. But, what it does indicate is that results corresponding to the revised mesh are not valid because a new Solution has not yet been run based upon the revised mesh. Thus, it is logical that existing Study results are not valid.

13. Click **[Close]** to exit the **What's Wrong** window.

To address these warnings and errors, a new Solution, based on the revised mesh must be run for this Study. To accomplish this, proceed as follows.

14. Right-click **High Quality Mesh-COARSE (-Default-)** and from the pull-down menu, select **Run**. Aside: Note the speed with which this solution is completed.

Briefly examine contents of the **Results** folder to observe maximum stress magnitudes shown on **Stress1 (-vonMises-)**, and **Sigma-X (-X normal-)**, and also the maximum displacement shown on **Displacement1 (-X disp-)**.

A plot of only **Sigma-X (-X normal-)** is included in Fig. 24. This figure shows the **COARSE** size **High** quality mesh used for the current study. Compare it to the **DEFAULT** size **Draft** quality mesh shown in Fig. 16. Which has more nodes and elements? Which is more accurate? These questions are answered at the conclusion of this example.

OBSERVATIONS:

- Both the vonMises stress and normal stress in the X-direction (σ_x) are larger for the **High** quality coarse size mesh than predicted by the **Draft** quality default size mesh.

- The magnitude of the X-displacement is also larger for the **High** quality coarse mesh than that predicted using a **Draft** quality default mesh size. Recall the statement that **High** quality elements are not as "stiff" as **Draft** quality elements. This lower mesh stiffness results in larger deflections as will be observed again later in this example.

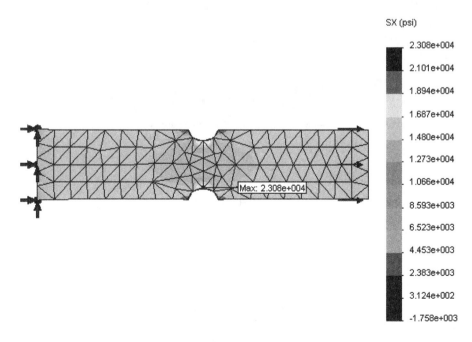

Figure 24 – A High quality, coarse mesh displayed on the Notched Bar model.

Study Using High Quality Elements and DEFAULT Mesh Size

In this section the Notched Bar example is solved again, but this time using a High quality **Default** size mesh (i.e., a smaller size mesh). Because this new solution is basically a repeat of the section above, complete, but abbreviated instructions are provided below. Practice performing many of these steps on your own.

1. In the COSMOSWorks analysis manager, click-and-drag the **High Quality Mesh-COARSE (-Default-)** folder onto **Notched Bar** located at the top of the COSMOSWorks manager tree. The **Define Study Name** window opens as shown in Fig. 25.

2. On the **Study Name:** line of this window, type: "**High Quality Mesh-DEFAULT**."

3. Click **[OK]** to close the **Define Study Name** window.

4. A new Study folder named **High Quality Mesh-DEFAULT (-Default-)** appears in the COSMOSWorks manager tree.

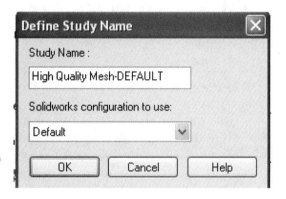

Figure 25 – Naming the third mesh study.

Next, alter the mesh size to match the name just assigned to this new study.

5. Right-click the **Mesh** folder and from the pull-down menu, select **Create Mesh...**. A COSMOSWorks warning window opens in Fig. 26 and displays the message: "**Remeshing will delete the results for study: High Quality Mesh-DEFAULT**."

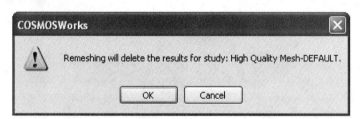

Figure 26 – Warning alerts the user that results associated with the current mesh will be deleted.

6. Click **[OK]** to delete the current results. The **Mesh** property manager opens.

7. In the **Mesh Parameters** dialogue box, click the **[Reset]** button. This resets mesh size to the default size (pointer moves to middle of the upper slide-scale and mesh size resets to 0.16846771 in).

Stress Concentration

8. Click ⌄ to open the **Options** dialogue box and verify that **Mesh Quality** is still set to ◉ **High**.

9. Click **[OK]** ✓ to close the **Mesh** property manager.

The model is re-meshed and once again an *Error* and a *Warning* appear in the COSMOSWorks manager. Eliminate them by running a new Solution as outlined below.

10. Right-click **High Quality Mesh-DEFAULT (-Default-)** and from the pull-down menu, select **Run**. Aside: Once again, note the speed with which this solution is completed. This solution should run slightly longer than the previous solution.

Briefly examine maximum magnitudes on plots of **Stress1 (-vonMises-)**, and **Sigma-X (-X normal-)**, and **Displacement1 (-X disp-)** in the **Results** folders. Because of the smaller mesh size, all values should *exceed* those of the previous Study.

Study Using High Quality Elements and FINE Mesh Size

Next, solve the Notched Bar example again, but this time use a Fine size **High** quality mesh. The solution process parallels that of the preceding two sections with the exception that a new descriptive name is assigned to the study (include the word "**FINE**" to identify the new mesh size) and reset mesh size to **Fine** when re-meshing the model. Because this new solution is a repeat of the previous two solutions, try to solve it without referring to these notes. However, if guidance is desired, follow the abbreviated procedure in the preceding section. Notice the significant increase of Solution time required. Complete this Study by examining results and comparing magnitudes with prior results.

Study Using High Quality Elements and MESHCONTROL

Perhaps the *most important* approach to refining mesh size is that of applying *mesh control*, which is investigated in this section. By this point in the example it is evident that a smaller and higher quality mesh results in improved approximations of stress in a model. However, the approach of progressively using smaller and smaller mesh size

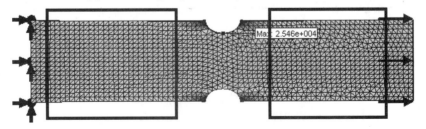

Figure 27 – "Boxed" areas denote regions where *inefficient* use is made of a **Fine** size mesh because stress is uniform (constant) or nearly uniform in these regions of the model.

throughout the *entire* model (i.e., use of a global mesh size) results in significant inefficiencies during the solution process. For example, when the **Fine** mesh of the previous section is used, consider the large number of equations solved in those regions of the model where stress is nearly uniform as illustrated by "boxed" regions in Fig. 27.

However, use of *mesh control* allows the user to control element size on selected entities of the model, such as **Faces**, **Edges**, **Vertices** or **Reference Points**. Controlling mesh size makes it possible to specify a smaller mesh in selected regions of high stress gradient, while simultaneously using a significantly larger mesh size in regions where stress distribution is relatively uniform. Locating local regions of high stress at the start of an analysis is easily accomplished by using a default mesh. This fact was observed earlier in this example when high stress magnitudes were found in the region of the "notch." The procedure below outlines steps for applying mesh control to a select part of the model.

As in the preceding section, the Notched Bar example is solved again. But, this time a High quality **Default** size mesh is used except in the region around the "notch." Because use of mesh control introduces new software capabilities not used above, all steps in this procedure are outlined below.

1. In the COSMOSWorks manager, click-and-drag the **High Quality Mesh DEFAULT (-Default-)** folder onto **Notched Bar** located at the top of the COSMOSWorks manager tree. The **Define Study Name** window opens as shown in Fig. 28.

2. On the **Study Name:** line, type: "**High Quality-MESH CONTROL**."

3. Click **[OK]** to close the window.

4. A new Study folder named **High Quality-MESH CONTROL (-Default-)** appears in the manager tree.

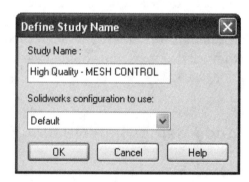

Figure 28 – Naming the fifth mesh Study.

The following steps introduce use of *mesh control*.

5. Right-click the **Mesh** folder and from the pull-down menu, select **Apply Mesh Control…**. The **Mesh Control** property manager opens as illustrated in Fig. 29.

As noted above, an advantage of using mesh control is that it can be applied to local regions on a model. In the following steps, portions of the model in the vicinity of the stress concentration are selected.

6. In the **Selected Entities** dialogue box, the **Faces, Edges, Vertices, Reference Points, Components for Mesh Control** field is highlighted (light blue) indicating it is active.

7. Move the cursor onto the model and *at both the top and bottom notches*, select the two curved **Edges** and one **Face** at each notch as illustrated in Fig. 30.

   Edge<1>, **Edge<2>**, **Edge<3>**, **Edge<4>**, and **Face<1>**, **Face<2>** appear in the highlighted field of Fig. 29 in the order selected by the user.

At this point, no changes are made to default values in the **Control Parameters** dialogue box. However, the three fields are explained below.

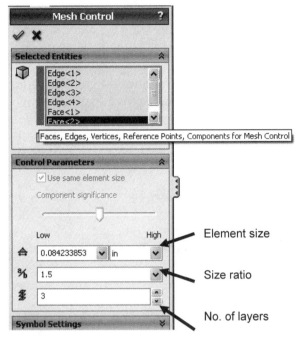

Figure 29 – Entities to which mesh control is applied and mesh size variation is specified in the **Mesh Control** property manager.

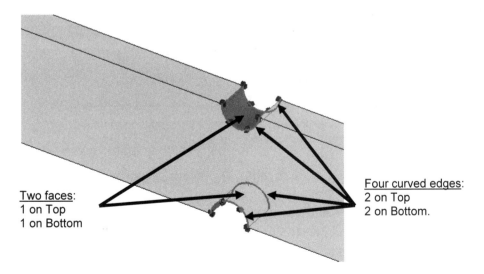

Figure 30 – Edges and faces selected for application of **mesh control** at top and bottom notches. Mesh control symbols are shown on selected entities.

3-23

Element Size: A system determined element size is applied to the selected entities. The initial element size (0.084233853 in) corresponds to the **Fine** mesh setting used earlier in this example.

Ratio a/b: This value controls the increase of element size from one layer to the next as elements radiate away from the selected entities.

Layers: This value defines the number of layers (in this case 3) used to gradually change from the small **Element Size** (defined above) to the global element size used for the remainder of the model.

Steps below complete the mesh definition.

8. Click **[OK]** ✓ to close the **Mesh Control** property manager.

In the COSMOSWorks manager tree, observe that error symbols appear adjacent to the **Mesh** icon and the Study folder, named **High Quality-MESH CONTROL (-Default-)**. A warning symbol also appears next to the **Results** folder. These errors and warning are due to the change of mesh definition. To verify this, right-click either folder and select **What's wrong?**.... After reading the message about mesh changes, click the **[Close]** button. Also notice that a **Mesh Controls** icon and a **Control-1** icon appear below the **Mesh** folder.

Proceed as follows to update the mesh and re-run the solution.

9. Right-click the **Mesh** folder and from the pull-down menu, select **Create Mesh…**. A **COSMOSWorks** warning window opens in Fig. 31.

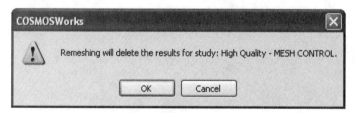

Figure 31 – Warning alerts the user that results associated with the current mesh will be deleted.

10. Click **[OK]** to delete the current results. The **Mesh** property manager opens.

11. In the **Mesh Parameters** dialogue box, click the **[Reset]** button. This action establishes the default mesh size for portions of the model away from entities where mesh control is applied.

12. Click **[OK]** ✓ to close the **Mesh** property manager.

The model is re-meshed and an enlarged view of the mesh in the vicinity of the notches is shown in Fig. 32.

In the COSMOSWorks manager tree the error symbol is now removed from the **Mesh** folder, but an error still remains adjacent to the Study name. Eliminate it by running a new Solution as outlined next.

13. Right-click **High Quality-MESH CONTROL (-Default-)** and from the pull-down menu, select **Run**. Alternatively, click the COSMOSWorks **Run** icon. Aside: Once again note the faster speed with which this solution is completed.

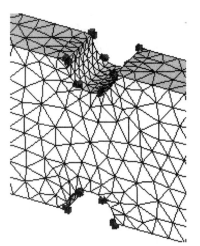

Figure 32 – View of mesh size altered by application of **Mesh Control** in the vicinity of the notch.

Briefly examine results shown on plots in the **Results** folder. Then proceed to gain greater insight into factors affecting **Mesh Control** by making the changes outlined below.

14. Right-click the **Control-1** icon beneath the **Mesh** folder in the COSMOSWorks manager tree and from the pull-down menu select **Edit Definition…**. The **Mesh Control** property manager opens as previously shown in Fig. 29.

15. In the **Control Parameters** dialogue box change the **Ratio** value to **1.2**. Also change the **Layers** value to **5**.

These changes decrease element size changes from one layer to the next while increasing the number of layers between the smallest elements and default size elements used for the remainder of the model.

16. Click **[OK]** ✓ to close the **Mesh Control** property manager.

17. Right-click the **Mesh** folder and from the pull-down menu, select **Create Mesh…** A **COSMOSWorks** warning window opens again as previously shown in Fig. 31.

18. Click **[OK]** to delete the current results. The **Mesh** property manager opens.

19. In the **Mesh Parameters** dialogue box, verify that the **Default** mesh size is still selected and click **[OK]** ✓ to close the **Mesh** property manager.

The model is re-meshed automatically and the new mesh is shown in Fig. 33. Visually compare differences of mesh size and the more gradual transition from a small near the notch to the default size mesh illustrated in Figs. 32 and 33.

20. Complete the solution by right-clicking **High Quality-MESH CONTROL (-Default-)** and select **Run** from the pull-down menu. Aside: Once again notice the faster speed with which this solution is completed.

By using *mesh control*, a small mesh is created in areas of high stress gradient while a larger mesh is used in other regions of the model. Results of the current study are recorded below and compared with results of all previous studies included in this example.

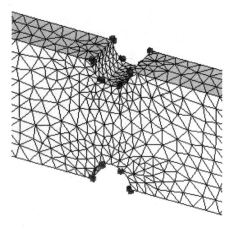

Figure 33 – **Mesh Control** using a smaller **Ratio** and higher number of **Layers** around the notch.

Results Analysis

Now that five different meshes have been applied to the Notched Bar, the significance of the resulting solutions is investigated and discussed below. Begin by creating a *simultaneous* display of results from the four analyses for which *global* element size was changed. To open multiple viewports, proceed as follows. *SolidWorks users may be familiar with this software feature and are encouraged to proceed on their own.*

Create Multiple Viewports

1. In the Main menu, at top of screen, click **Window**. Figure 34 shows the pull-down menu and the options available for displaying multiple window configurations.

2. Move the cursor onto **Viewport** and from the second-level pull-down menu, select **Four View**.

The graphics area immediately displays four views of the current screen image. This image is altered in the following steps.

Various plots can be displayed in these viewports in keeping with user preferences. However, to facilitate discussion, it is suggested that the following order be used. Begin by placing a plot of vonMises stress found using the **Draft** quality mesh in the upper left-hand viewport. The following steps outline this procedure.

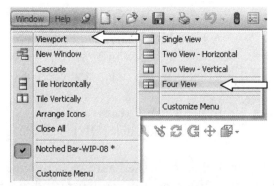

Figure 34 – Menu selections used to open multiple viewports.

Stress Concentration

3. Click anywhere in the upper left-hand viewport (labeled "**A**" in Fig. 35). This action activates the viewport.

4. In the COSMOSWorks manager, click the "+" sign adjacent to **Draft Quality Mesh DEFAULT (-Default-)**. This action displays a listing of the Study contents in the COSMOSWorks manager tree.

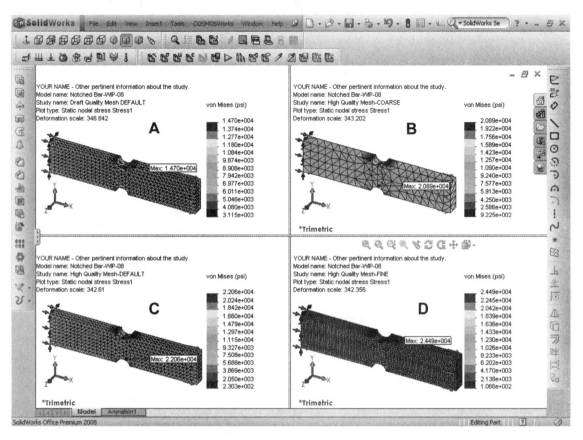

Figure 35 – Multiple viewports facilitate comparison of results from multiple different studies or different aspects of a model within a single study.

5. Next, click "+" adjacent to the **Results** folder to display its contents.

6. Finally double-click **Stress1 (-vonMises-)** and the contents of that plot are displayed in viewport "**A**" as illustrated in Fig. 35. See NOTES below.

NOTE 1: Do not be concerned if your *view* of the model differs from that shown. Desired views can be adjusted later based upon user preferences. In fact, notice that the default view orientation corresponds to Top, Isometric, Right-side, and Front views of
the model in a clockwise direction beginning at the top-left of the screen.

NOTE 2: Viewport labels **A**, **B**, **C**, **D** are added to the textbook image to facilitate discussion. They do not appear on the COSMOSWorks screen image.

3-27

7. Click in viewport **B** and repeat steps 4 to 6 for the von Mises stress plot of the **High Quality Mesh-COARSE (-Default-)**.

8. Click in viewport **C** and repeat steps 4 to 6 for the von Mises stress plot of the **High Quality Mesh-DEFAULT (-Default-)**.

9. Finally, click in viewport **D** and repeat steps 4 to 6 for the von Mises stress plot of the **High Quality Mesh-FINE (-Default-)** study. This display takes longest to appear due to more data associated with the fine mesh.

Now that results of the four global element size studies are displayed in individual viewports, screen images can be adjusted according to user preferences. Click in the viewport to be modified and use traditional pan, zoom, rotate and view options to adjust the image to best display results. Typically the choice of view depends upon those aspects that are of primary importance to a study. For example, a close-up view of the notch might be used to replace one, or all, of the views shown above. Trimetric views of each meshed model are shown in Fig. 35. To obtain these views, click inside each viewport and select the Trimetric icon. This view is selected because it provides a good overview of mesh size and stress magnitudes that are important to this analysis.

Your graphics screen should look *similar* to Fig. 35. The left side of the graphics screen is minimized here to focus attention on results. *Close this study without saving results.*

What Can Be Learned From This Example?

Most of the primary goals of this analysis were encountered and mastered as you worked through the example itself. Those goals included: knowing 'when,' 'how,' and 'why' a model should be *defeatured*; using the *copy* feature to quickly create new plots and new studies; using multiple *viewports* to facilitate results comparisons; and, most important, using *global mesh refinement* and *local mesh control* as analysis tools. The importance of each capability as part of an overall Finite Element Analysis (FEA) is discussed further in the "Analysis Insight" section below.

Other Uses of the Copy Feature

The *copy* feature within COSMOSWorks is not limited to creating new studies in which changes to mesh size is the only analysis variable. In fact, any variable of a finite element analysis can be the central focus in a multiple solution study. For example, other variables that could be altered during a repetitive analysis include, but are not limited to: material properties, loads, restraints, or modifications to the model geometry itself.

Analysis Insight

The importance of mesh refinement and mesh control as part of a complete finite element analysis cannot be overstated! A single solution to a finite element analysis provides only one snapshot of what a possible solution to a problem might be. A primary reason for performing multiple analyses is to determine whether or not a study is converging to an acceptable solution.

Analysis Insight (continued)

To determine *convergence*, it is possible to examine either stress results or displacement results. From the Introduction to this text, recall that displacements are the primary unknowns in a finite element analysis. From displacements the strains are calculated, and finally, from the strains the stresses are computed. Because displacements are the first link in the solution chain, they are usually a better indicator of convergence to a solution as element size is altered. For the sake of brevity, only vonMises stress results and X-displacement results corresponding to *global* mesh size variation are compared below when testing for convergence.

To relate this discussion to the Notched Bar, **Table 1** summarizes results of all analyses developed in this example. Von Mises stress data and displacement data in **Table 1** are obtained from Fig. 35 and mesh control plots created throughout this study. Information about the number of nodes and elements in each mesh is obtained by right-clicking each **Mesh** folder and selecting **Details....** This procedure was demonstrated in Chapter 1.

In **Table 1**, focus your attention on von Mises stress results column corresponding to the **High** quality mesh contained in the middle three rows of the table only. These three rows show how stress magnitude increases as mesh size varies from 'coarse' to a 'fine'. Casual observation of magnitudes listed in the von Mises stress column reveals an increase of stress as the number of elements increases. This result is expected due to the ability of smaller elements to better delineate stress magnitudes in increasingly smaller regions of high stress gradient. This finding might lead one to expect that as element size gets smaller and smaller, the stress values will continue to grow larger and larger. However, in a properly constructed study quite the opposite is true! In an analysis that is *converging* to a solution, stress results tend to *level off* to some limiting value where that limiting value should represent the best approximation to stress in the model. This is the reason that *multiple* solutions should be a standard part of every finite element analysis (i.e., to determine *convergence* to a solution).

TABLE 1 – Summary of Results from the Notched-Bar Study

Mesh	No. Nodes	No. Elements	vonMises Stress (psi)	Sigma-X (psi)	Displacement X-Direction (in)
DRAFT Quality (Default size)	1,603	6,437	14,700	15,740	0.001874
HIGH Quality (Coarse size)	1,904	1,029	20,890	23.080	0.001894
HIGH Quality (Default size)	10,614	6,437	22,060	23,650	0.001897
HIGH Quality (Fine size)	81,287	55,008	24,490	25,460	0.001898
HIGH Quality (Mesh Control)	15,428	9,669	24,180	25,420	0.001898

Analysis Insight (continued)

If the von Mises stress values are plotted against an increasing number of elements, the graph in Fig. 36 is obtained. Results depicted in Fig. 36 show a slight, as opposed to a rapid, increase of stress corresponding to a doubling of the number of elements from point to point along the abscissa. This trend, while not totally definitive, indicates stress values may be reaching a limiting value.

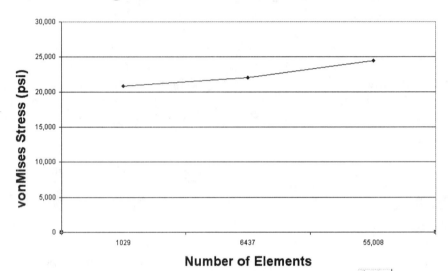

Figure 36 – Convergence check using vonMises stress as a measure.

Displacement versus the number of elements is examined next to determine whether or not it provides a better indication of convergence to a solution than does the von Mises stress. Data in the displacement column of **Table 1** reveals an upward trend as the number of elements increases (i.e., as global mesh size gets smaller).

When plotted, displacement versus number of elements yields the graph shown in Fig. 37. In addition to straight line segments drawn through the data points, a higher-order curve fit was applied to the data resulting in the curved line shown. In either case, the "leveling off" of results (i.e., convergence to a maximum displacement value) is indicative that the solution is converging.

Stress Concentration

Analysis Insight (continued)

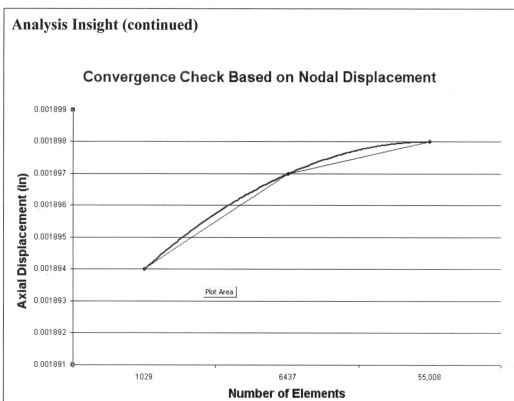

Figure 37 – Convergence check using displacement as a measure.

Summary:
The *need* to conduct more than a single finite element analysis is demonstrated in this example. The *reason* for conducting more than one analysis is to determine whether or not the analysis is approaching a limiting value. This is known as a convergence check.

Aside #1:
Results of the **Draft** quality mesh are *excluded* from the above discussion because it is most revealing to compare results obtained using the same *type* of mesh. All other meshes were **High** quality.

Aside #2:
Comparison of displacement results for **Draft** quality and **High** quality meshes of the *same* size, shown in **Table 1**, confirms that a **Draft** quality mesh in stiffer than a **High** quality mesh because a smaller displacement results for the **Draft** quality mesh.

Aside #3:
Despite the emphasis on *global* mesh size refinement in this example, perhaps the most important means for controlling mesh size is use of *mesh control*. Mesh control can be applied on a *local* basis and provides accurate results at significantly lower computational overhead. For example, comparing values in **Table 1**, the mesh control approach uses approximately five times fewer nodes and elements than does a globally **Fine** mesh, yet von Mises stress results differ by less than 0.16%.

Comparison of Classical and FEA Results

The importance of checking finite element results is emphasized throughout this text. Therefore, in keeping with this premise, it is appropriate to conclude this example with an analytical check. Figure 2 is repeated below.

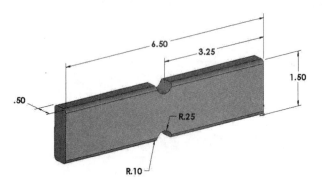

R = notch radius = 0.25 in.

h = height (top to bottom) = 1.50 in

d = h − 2*R = 1.00 in (dimension at the reduced cross-section)

t = bar thickness = 0.50 in

F = applied force = 6000 lb

Figure 2 (repeated) – Dimensioned image of notched bar.

Nominal stress at the reduced cross-section in the Notched Bar:

$$\sigma = \frac{F}{A} = \frac{F}{d*t} = \frac{6000}{1*0.5} = 12000\ psi$$

Geometric stress concentration factor: where: R/d = 0.25/1.00 = 0.25
  h/d = 1.50/1.00 = 1.50

K_t = 2.1 (approximate) ← From stress concentration factor chart found in a standard machine design textbook.

Maximum stress in the Notched Bar:

$\sigma_{max} = K_t * \sigma$ = 2.1 * 12000 = 25200 psi

Percent difference:

% Difference = [(COSMOSWorks − Classic)/ (COSMOSWorks)] * 100

% Difference = [(25460 − 25200)/(25460] * 100 = 1.02%

Note that Sigma-X (σ_x) is used for comparison purposes since the classical equations for stress concentration depend upon axial stress as opposed to vonMises stress. The above calculations provide a good check of finite element results. Note that resultant stress (σ_x)$_{Max}$ predicted by the mesh control model yields 0.87 % difference.

EXERCISES

End of chapter exercises are intended to provide additional practice using principles introduced in the current chapter plus capabilities mastered in preceding chapters. Most exercises include multiple parts. Maximum benefit is realized by working all parts. However, in an academic setting, it is likely that parts of problems will be assigned or modified to suit specific course goals.

1. A rectangular bar with a centrally drilled hole is illustrated in Fig. E 3-1. The bar is supported **(Immovable)** at its left-end and subject to an axial, tensile force of 370 kN applied normal to its opposite end. The bar is made from 2018 aluminum alloy. Open the file: **Plate With Hole 3-1**.

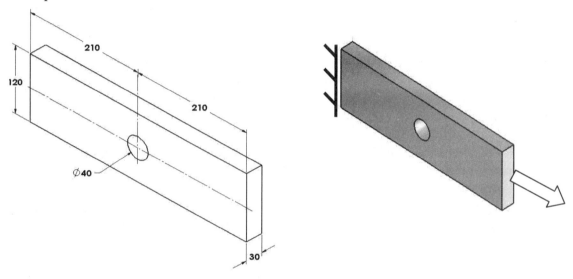

Figure E 3-1 – Aluminum bar with central hole subject to an axial force. A geometric discontinuity is present in the form of the 40 mm diameter hole.

- Material: **2018 Alloy** aluminum (Use S.I. units)

- Mesh: **High Quality** tetrahedral elements. Use three different meshes as specified in parts (b, c, and d) below.

- Restraint: **Immovable** applied to left end.

- Force: **370 kN** applied normal to the right-end causing tension in the bar.

Develop a finite element model that includes: material specification, restraints, applied force, mesh generation, and solution as specified below. For this analysis use a default size, high quality mesh and perform the following:

Analysis of Machine Elements using COSMOSWorks

Determine the following:

a. Using a *default* size mesh, create a contour plot of the *most appropriate stress* to permit comparison of its magnitude adjacent to the hole with that predicted by classical equations for stress computed at the same location; see part (e) for calculation of classical results. Include the restraints, applied load, and mesh on this plot.

b. Use the **Probe** feature to produce a graph of the *most appropriate stress* from the top (or bottom) edge of the bar to the edge of the central hole. Because a straight path may not be available, choose both corner and mid-side nodes in as straight a line as possible.

c. Repeat part (b) after resetting the mesh size to *fine*. Use the copy feature to save time creating this study.

d. Repeat part (b) a third time after resetting the mesh size by applying *mesh control* around both edges and the inner surface of the hole. Use the default mesh control setting: **Ratio a/b = 1.5** and **Layers = 3**. Also use the copy feature to save time creating this study.

e. Use classical equations and available stress concentration factor charts to compute maximum stress at the hole.

f. Compare results predicted using the three different meshes with that predicted by classical stress equations. Compute the percent difference for each comparison using equation [1].

$$\% \text{ difference} = \frac{(\text{FEA result - classical result})}{\text{FEA result}} * 100 = \qquad [1]$$

g. Comment upon which FEA results are in best agreement with predictions of the classical equations? Which method of mesh refinement is usually preferred and why?

h. Based upon results for the maximum *appropriate stress* in the part, is the material Yield Strength exceeded? If "yes," what does theory predict will be the outcome in a ductile material such as aluminum used in this example?

2. A cylindrical rod changes diameter at the fillet shown in the Fig. E3-2. It is well known that generous fillets are beneficial in reducing stress concentration at changes of cross-section. This is particularly important because the bar in question is made from a relatively brittle material, gray cast iron, and the fillet radius is not very large. As such, the rod is subject to significant effects of stress concentration at the

geometric discontinuity. Consider the rod supported **(Immovable)** at its left-end and subject to a 1,200 lb tensile force applied to its opposite end.

Open the file: **Shaft with Fillet 3-2**.

- Material: **Gray Cast Iron** (Use English units)

- Mesh: **High Quality** tetrahedral elements. Use three different meshes as specified below.

- Restraint: **Immovable** applied to left end.

- Force: 1,200 lb applied normal to the right-end causing tension in the bar.

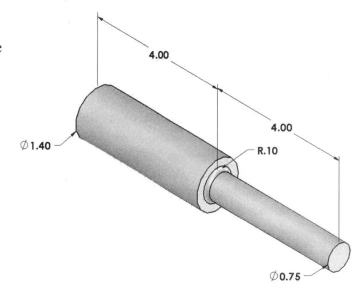

Figure E3-2 – Gray Cast Iron cylindrical rod subject to an axial tensile load of 1,200 lb.

Determine the following:
a. Develop a finite element model that includes: material specification, restraints, applied force, mesh generation (as specified below) and solution.

NOTE: This example is one where it would *not* be appropriate to defeature the model because deleting the fillet radius would cause a dramatic increase of stress at the change of shaft diameter. In fact, a sharp fillet would probably result in stress values continuing to increase as the mesh size gets smaller. This is an example of a solution where *divergence* from a solution rather than *convergence* to a solution would occur.

b. Perform part (a) using a default size, high quality mesh. For this mesh, perform the following:

- Create a stress contour plot of the *most appropriate stress* to permit comparison of stress magnitude in the vicinity of the fillet with that predicted by classical stress equations computed at the same location. Include a mesh on this plot.

- Use the **Probe** feature to produce a graph of the *most appropriate stress* commencing approximately ½ in. to the right of the fillet and progressing, in as near a straight-line as possible, through the fillet area up to the outside diameter of the left segment (large diameter segment) of the shaft. Choose both corner and mid-side nodes.

c. Repeat part (b) after resetting the mesh size to fine. Use the copy feature to save time creating this study.

d. Repeat part (b) again beginning with the default size mesh.. Use the copy feature to save time creating this study. Then, alter the mesh by applying mesh control on the fillet. Use the default mesh control setting: **Ratio a/b = 1.5** and **Layers = 3**.

e. Use classical equations and available stress concentration factor charts to compute maximum stress at the fillet.

f. Compare results predicted using the three different meshes with that predicted by classical stress equations; clearly label each calculation. Compute the percent difference for each comparison using equation [1] (repeated).

$$\% \text{ difference} = \frac{(\text{FEA result - classical result})}{\text{FEA result}} * 100 = \qquad [1]$$

g. Determine the Safety Factor, or lack thereof at the change of cross-section (i.e., at the fillet). Create a plot showing regions (if any) of the model with Safety Factor less than one.

h. Comment upon which FEA results are in best agreement with predictions of the classical equations? Based on your comparison, which method of mesh refinement is preferred and why?

i. For users conversant with modifying solid models in SolidWorks, proceed to alter the fillet radius at the change of cross-section to a radius equal to the difference of the two shaft sizes and repeat parts assigned by your instructor. Report the amount by which it is possible to reduce stress at the change of cross-section.

Textbook Problems
In addition to the above exercises, it is highly recommended that additional problems involving stress concentration and/or significant changes of part geometry be worked from a design of machine elements or mechanics of materials textbook. Textbook problems provide a great way to discover errors made in formulating a finite element analysis because they typically are well defined problems for which the solution is known. Typical textbook problems, if well defined in advance, make an excellent source of solutions for comparison.

CHAPTER #4

THIN AND THICK WALL PRESSURE VESSELS

This chapter investigates modeling of thin and thick wall pressure vessels. More importantly, however, the use of *shell* elements is introduced and guidelines are provided regarding their use as opposed to *solid* tetrahedral elements. When taken alone, thin and thick wall pressure vessel problems are solvable in rather a straight-forward manner by using classical stress equations. Therefore, advantages of the Finite Element Analysis (FEA) approach are realized most when investigating stresses in more geometrically complex regions of a pressure vessel. These regions include, but are not limited to, locations of pipe connections, saddles, or other flanges or support structures associated with pressure vessel design and installation. Many of these special situations can be handled by application of principles outlined in preceding chapters. Thus, the two examples of this chapter focus on introducing additional capabilities of the finite element software rather than on pressure vessels per se'.

Learning Objectives
Upon completion of this example, users should be able to:

- Recognize conditions that favor use of *shell* elements, as opposed to solid elements, and use a *shell* mesh to model thin-wall parts (whether or not the part in question is a pressure vessel).

- Recognize part *symmetry* and use it to reduce solution size and computation time. Know when and how to apply *symmetrical restraints*.

- Be able to apply *section clipping* to enhance viewing of results.

- Apply uniform *pressure* loading.

THIN-WALL PRESSURE VESSEL

Thin-wall pressure vessels are found in many common applications. Included among them are carbonated beverage containers (aluminum cans and plastic bottles), aerosol cans used to dispense everything from hair spray to insecticides, hydraulic cylinders, steam boilers, and tall water-tower tanks that provide pressure to public water systems.

Problem Statement
This example is based on a thin-wall cylindrical pressure vessel closed on both ends with hemispherical heads as illustrated in Fig. 1. Wall thickness t = 3 mm and inside diameter of the cylinder d_i = 144 mm. The vessel is made of **AISI 1045** cold drawn steel and is subject to an internal pressure **P = 1.4 MPa**. Other dimensions are included in Fig. 1.

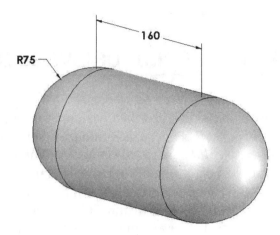

Figure 1 – Basic dimensions of a thin-wall pressure vessel closed with hemispherical ends.

The dividing line between thin and thick wall pressure vessels is defined differently in various machine design[1] and mechanics of materials texts. If wall thickness "t" falls in the range where $t \leq r_i/20$ to $t \leq r_i/5$, (where, r_i = cylinder inside radius), then the cylinder is considered a thin-wall pressure vessel. This determination must be made at the outset of any pressure vessel problem because choice of the appropriate set of classical equations depends upon whether wall thickness is classified as "thin" or "thick." By the first definition above, the current model can be considered a thin-wall pressure vessel since the minimum criteria yields: $t = 3$ mm $\leq (72$ mm$)/20$. Another important aspect of thin-wall pressure vessels is that magnitude of the tangential stress (σ_t), also known as "hoop" stress or "circumferential" stress, is assumed *uniform* through the wall thickness.

Primary stresses in the pressure vessel walls are given by:

Tangential stress: $$\sigma_t = \frac{p*d_i}{2*t} = \frac{(1.4e6 \frac{N}{m^2})(0.144 m)}{2(0.003 m)} = 33.6e6 \text{ Pa} \quad [1]$$

Longitudinal stress: $$\sigma_\ell = \frac{p*d_i}{4*t} = \frac{1}{2}\sigma_t = \frac{1}{2}(33.6e6 \text{ Pa}) = 16.8e6 \text{ Pa} \quad [2]$$

(for a closed-end cylinder)

Where: p = pressure (N/m^2) = 1.4 MPa
   t = wall thickness (m) = 0.003 m
   d_i = inside diameter (m) = 0.150 m – 2*0.003 m = 0.144 m

These two stresses are perpendicular to sides of a stress element aligned with and perpendicular to the longitudinal axis of the cylinder shown in Fig. 2. Because no shear stresses act on sides of the element shown in Fig. 2, σ_t and σ_ℓ are principal stresses.

---

[1] Budynas, R.G., Nisbett, J.K., Shigley's Mechanical Engineering Design, use: $t \leq r_i/20$.
Collins, J.A., Mechanical Design of Machine Elements, uses: $t \leq 10\%*d = (.1)(2*r) = r_i/5$.

Finally, all stresses on the hemispherical ends are tangential stresses, but their magnitude is half of that in the cylindrical portion of the vessel.

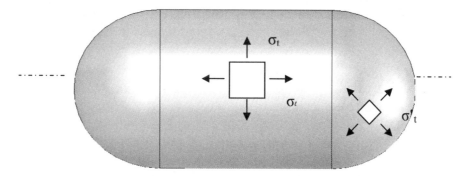

Figure 2 – Stress elements showing orientation of σ_t and σ_l on the pressure vessel surface and on the hemispherical end. ($\sigma'_t = \frac{1}{2}\sigma_t$)

Due to St. Venant's principle, normal stresses depicted in Fig. 2 are only valid at locations well removed from the junction between the cylindrical segment and the hemispherical ends where end-conditions exist. Thus, to permit comparison with results of equations [1] and [2], the following finite element analysis focuses on stresses near the mid-section of the cylinder and on the hemispherical heads.

The next decision to be made is whether to use *solid* or *shell* elements for the finite element model. The guideline for this decision is quite different than that applied above to determine what set of classical equations should be used. For the current model, it is possible to use either *solid* or *shell* elements. Both would produce valid results. However, the selection of element type has a significant influence on the number of elements in a model and hence upon solution time and memory requirements.

Shell elements are typically used for sheet metal and other thin parts. The guiding principle used to decide what element type to use is that *solid* elements are applied when two (default size) elements nicely fit across the thickness of a part (i.e., across the part's minimum dimension). Although element size can be made sufficiently small to fit two elements across the 3 mm wall thickness, that small size results in an extremely large number of elements on the entire model. For this reason, *shell* elements, which are only one "layer" thick, are selected.

Shell elements, which assume the shapes pictured in Figs. 6 and 7 of the **Introduction**, are repeated below in Figs. 3 (a) and (b).

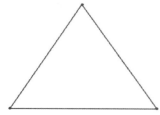

Figure 3 (a) – First-order triangular shell element.

Figure 3 (b) – Second-order shell element includes three additional mid-side nodes.

Analysis of Machine Elements using COSMOSWorks

Shell elements look somewhat similar to solid elements except they are represented graphically as a single layer of elements, as if they were drawn on a thin sheet of paper. Later images will illustrate shell elements. However, because parts modeled using shell elements must have a finite thickness, a corresponding thickness is assigned to shell elements as described later. The finite element analysis of a pressure vessel using shell elements is outlined below.

1. Open SolidWorks by making the following selections. (*Note:* "/" is used to separate successive menu selections.)

    Start / All Programs / SolidWorks 2008

2. From the main menu within SolidWorks select **File / Open**. Then proceed to where your files are saved and open the COSMOSWorks file named **Thin Wall Pressure Vessel**. *It is strongly suggested that this model file be used for the current example rather than a user created pressure vessel model.*

Understanding System Default Settings

Because earlier examples required frequent specification of items such as mesh quality, units, or other attributes of an analysis, and because most of these items are defined as *system default settings*, a time savings in future examples will be realized by digressing to investigate these settings now.

1. Click the **COSMOSWorks** icon to open the COSMOSWorks manager.

2. If a SolidWorks window opens and indicates the **The following documents will be converted when saved: Thin Wall Pressure Vessel.SLDPRT**, click **[OK]**.

3. In the COSMOSWorks manager, right-click **Thin Wall Pressure Vessel** and from the pull-down menu select **Options…** The **Systems Options – General** window opens as shown in Fig. 4.

Alternatively, click **COSMOSWorks** in the main menu and from the pull-down menu select **Options…**. Either action opens the **Systems Options - General** window.

In this window, the **Systems Options** tab is initially selected as illustrated in Fig. 4. In the list at left of this window, the word General is highlighted to denote what settings are currently displayed. Because the current study focuses on use of a shell mesh, in the right half of this window, observe the default color scheme used to denote the bottom surface of shell elements. The text **Shell bottom face color** appears next to the system default color, orange, indicated adjacent to the arrow in Fig. 4. Before proceeding, observe other system default settings in this window. In particular, note that ☑ **Show yield strength marker for vonMises plots** is checked. Recall that this feature was used in the curved beam analysis of Chapter 2.

Pressure Vessels

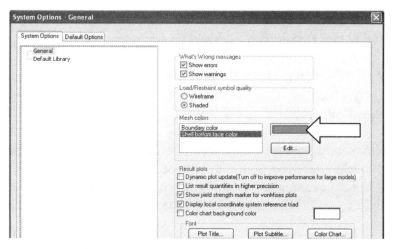

Figure 4 – Default settings beneath the **System Options** tab. Default color of the bottom surface of a shell mesh is shown here.

4. Next click the **Default Options** tab. Within this tab Units is highlighted (gray) to indicate it is selected in the list at the upper-left of the window. Only the right-half of this window is illustrated in Fig. 5. Within this window, most of the settings appear as listed below because SI units are default within COSMOSWorks; if not, change to the settings shown.

- Beneath **Unit System**, set default units to ⊙ **SI [MKS]**, if not already selected.

- Beneath **Units**, the units of **Length/Displacement**: should be set to **m** (meters).

- Adjacent to **Pressure/Stress:**, default units should be set to **N/m^2**. Other units do not pertain to this analysis and can be ignored.

Figure 5 – Selecting units to be applied throughout the analysis.

Specifying units as outlined above establishes a common set of units for an entire analysis. If, for example, your company or future employer uses English units exclusively, it would be a time-saver to reset default units accordingly.

5. Return to the list of **Default Options** tab at the left of this window and click to select Load/Restraint, the **Default Options – Load/Restraint** window is displayed.

Brief examination of the right-half of the window (not shown) reveals that the **Symbol size** and **Symbol colors** for various types of loads and restraints can be specified here.

4-5

Analysis of Machine Elements using COSMOSWorks

Although nothing is changed in this window, color-blind users may find it convenient to alter default **Symbol colors** to suit their personal preferences.

6. Next, from the list of **Default Options** at the left of the window, click to select **Mesh** and verify that the system default settings in the right half of the window appear as listed below.

Knowing the system default settings shown in this window and understanding that they remain unchanged unless altered by you, makes unnecessary many of the often repeated steps included in previous examples. For this reason, a brief explanation of each setting is included below. Reset any different settings to the default values listed below.

- **Mesh Quality** – set to ⊙ **High**. *Reason:* A high quality mesh is recommended for most analyses and for models with curved surfaces. A draft quality mesh is typically only used for an initial study where quick, approximate results are adequate.

- **Mesh Control** – clear the check mark from ☐ **Automatic transition**. *Reason:* This option automatically reduces element size at *every* change of geometry (fillets, rounds, holes, etc.). While this feature is nice, it can unnecessarily complicate complex models by adding numerous nodes and elements in regions of little or no interest.

- **Mesher Type** – clear the check mark from ☐ **Alternate**. *Reason:* This mesher takes longer to run because it automatically creates a smaller mesh in areas of high curvature. This mesher also requires setting incompatible bonding for assemblies and for situations where contacting features are involved. The default **Standard** mesher is faster and produces good results.

- **Jacobian Check** – set to ☑ **Solids** and ⊙ **4 points** selected. *Reason:* This option sets the number of integration points used when checking distortion levels in tetrahedral elements. The **4 points** selection is typically adequate for most analyses. Although **4** is shown as the standard, **6** points are automatically used for a shell mesh. Thus no change is necessary.

- Finally, ☑ **Automatic shell surface re-alignment** – should be selected. *Reason:* This option automatically ensures that adjacent surfaces of shell models are aligned during the meshing process.

Ignore remaining options within this window.

7. Next, from the **Default Options** list, select **Results**. Verify that the **Default solver** is set as ⊙ **FFEPlus**. *Reason:* **FFEPlus** is the fastest solver available in COSMOSWorks.

Pressure Vessels

8. Other options in this window direct analysis results to system default folders within SolidWorks. Unless other guidelines for file storage exist within your organization (university or company), use the default settings and ignore other settings in this window.

9. Return to the **Default Options** list and select **Plot**. The **Default Options - Plot** window opens. The right-half of this window is shown in Fig. 6.

10. Within this window, set **Fringe options:** to **Discrete**. This is a user preference, but is typically selected as **Discrete** in this text due to enhanced image characteristics when printed in black, white and gray tones.

11. Adjacent to **Boundary options:** select **Model** to show a black outline of the model. *Reason:* This option helps identify model boundaries when stress contour plots are viewed. This selection is also a user preference.

12. Another user preference is whether or not to display results on the deformed or undeformed shape of the model. Although this text typically elects display of the undeformed model, accept the default choice ⊙ **Show results on deformed shape**. Changes will be made on a case by case basis.

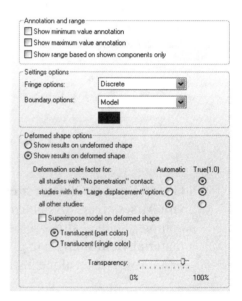

Figure 6 – The **Plot** window is used to specify display options for all plots.

Familiarize yourself with other settings within this window. Other settings should remain set to system defaults.

13. Next, in the **Default Options** list, click **Color Chart** and briefly observe items the user can control regarding display characteristics of the color chart including its placement on the screen, its size, and the format of numbers listed adjacent to the chart. Make no changes to default settings.

A nice feature about using the above capabilities is that each and every plot in the current and future studies can be produced with the same set of user selected settings. Pre-setting these parameters for all plots can save considerable time when a consistent set of plot characteristics is desired. However, using these capabilities does not preclude a user from altering characteristics of a particular plot at any point during a Study.

We next investigate default settings of all plots created during a Study.

Analysis of Machine Elements using COSMOSWorks

14. Click the "+" sign adjacent to **Default Plots** (if not already selected) and a listing of system default plots appears as shown on the left-side of Fig. 7.

15. At the left of the screen, below the **Static Study Results** folder, click to highlight the *name* **Plot1**. As of this writing, selecting the icon adjacent to **Plot1** does not activate this selection. The screen image shown in Fig. 7 appears.

16. In the right-half of the window, beneath **Results type:** select **Nodal Stress** (if not already selected). This selection is most typical because it gives stress magnitudes at specific locations (i.e., at each node) on the model.

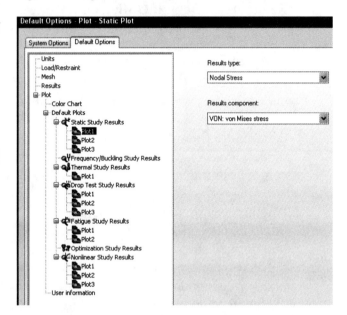

Figure 7 – Individual plots are pre-defined to indicate items and components to be plotted.

17. Observe that the default stress plot listed beneath **Results component:** is **VON: von Mises stress**. As noted in Chapter 2, von Mises stress represents the most complex state of stress as a single value. As such, its value is easily compared to material yield strength to give an indication of the safety (or lack thereof) in a part. This reason alone justifies it as worthy of being designated a default plot.

18. Return to the list of plots at the left of the screen and click the *name* **Plot2**. In the right-half of the screen observe **Displacement** as the **Results type:** and **URES: Resultant Displacement** for the **Results component:**.

19. Finally, click **Plot3** and observe the default definition of this plot is **Elemental Strain** as the **Results type:** and **ESTRN: Equivalent Strain** in the **Results component:** field.

Contents of the three default plot folders are based on the assumption that most users are interested in magnitudes of von Mises stress, model displacement and strain in the model. Although this is a logical assumption, this text focuses primarily on stress analysis.

If additional default plots are desired, they can be added now or in the future.
To demonstrate how to add an additional default plot, proceed as follows.

20. Right-click **Static Study Results** and from the pop-up menu, select **Add New Plot** as shown in Fig. 8. Immediately **Plot4** is added to the bottom of the current list.

    Define the contents of this new plot as follows.

21. In the right-side of the window, beneath **Results type:**, select **Nodal Stress** from the pull-down menu (if not already selected).

22. Beneath **Results component:**, select **P1: 1st Principal Stress**.

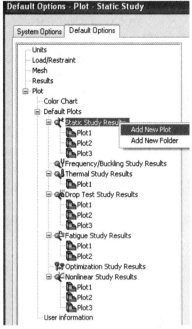

Figure 8 – Adding a new plot and including user information on all printed plots.

OBSERVATION:
Because the current Study involves analysis of stresses in the walls of a cylindrical pressure vessel, it is known in advance that tangential and longitudinal stresses will result. Further, it is well known that tangential stress (σ_t) and longitudinal stress (σ_l) correspond to the first and second principal stresses (σ_1 and σ_2) respectively. Therefore, *if* a user were dealing with pressure vessel analysis on a routine basis, it might make sense to add plots of the first and second principal stresses to the *default* set of plots produced at the conclusion of every study. However, since that is not the case for future examples included in this text, **Plot4** is deleted in the following step.

---

23. Right-click **Plot4** and from the pop-up menu, select **Delete**.

The above overview should provide sufficient insight so that users can access the **Default Options** window at the start of a Study to define those characteristics they wish to apply to a study. This action can save considerable time when common settings are used throughout a study.

24. Click the **[OK]** button to close the **Default Options** window.

Creating a Static Analysis using Shell Elements

1. Right-click **Thin Wall Pressure Vessel** at top of the COSMOSWorks manager tree and select **Study…** The **Study** property manager opens as shown in Fig. 9.

2. In the **Name** dialogue box, type **Pressure Vessel-Shell Study**.

3. In the **Mesh type** field, open the pull-down menu and select **Shell Mesh Using Mid-surfaces**.

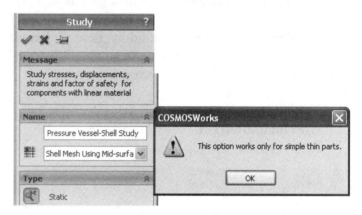

Figure 9 – **Study** property manager showing Study name, mesh type selection, and COSMOSWorks warning window.

A COSMOSWorks warning window opens, Fig. 9, and states, "**This option** (i.e., use of *shell* elements) **works only for simple thin parts**." This warning is important because it alerts the user not to use shell elements for other than simple, thin parts.

4. Click **[OK]** to close the **COSMOSWorks** warning window.

5. Click **[OK]** ✓ to close the **Study** property manager.

Assigning Material Properties

1. In the COSMOSWorks manager tree, right-click the **Solids** icon and from the pull-down menu select **Apply Material to All…**. The **Material** window opens.

2. Beneath **Select material source**, select ⊙ **From library files**.

3. Verify that **cosmos materials** appears in the selection box and click the "+" sign adjacent to **Steel (30)**.

4. From the list of 30 available steels, select **AISI 1045 Steel, cold drawn**. Before closing the **Material** window, beneath the **Properties** tab, notice that **Units:** is set as **SI** and corresponding units appear in the accompanying table. Recall that this selection of units was made at the beginning of this example.

5. Click **[OK]** to close the **Material** window.

Pressure Vessels

Assigning Loads and Restraints

Symmetry Restraints Applied

The thin-wall, cylindrical pressure vessel pictured in Figs. 1 and 2, is clearly a theoretical, "textbook" model. The rationale for this statement is quite obvious because no visible means of support are shown for the model as are no openings to permit fluids to enter or exit the container. However, this somewhat artificial geometry permits examination of methods to treat *symmetry* when it occurs in a problem.

Because the vessel geometry is symmetrical *and* because the internal pressure is uniform (symmetric loading), it is possible to analyze only a portion of the model. Figure 10 reveals that the current model was intentionally centered about the coordinate system origin when it was built. The front, right and top sketch planes, included in Fig. 10, aid in recognizing that the model can be divided into eight symmetrical pieces. Model symmetry can be used to save computer memory and computation time. Because this is a rather simple model, significant savings are not realized. However, consider the computational savings realized if symmetry is used to model a complete aircraft fuselage (a pressure vessel when in flight). Thus, this example demonstrates how symmetry can and should be used in other finite element solutions. Also, *symmetry* boundary conditions are introduced below.

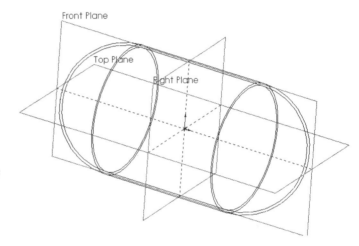

Figure 10 – Front, right, and top sketch planes introduced to show division of the model into symmetrical portions.

1. Click the **SolidWorks** icon at top-left of the feature manager tree to reveal its contents as shown in Fig. 11.

2. Next, right-click the grayed-out **1/4 Symmetry Model** near the bottom of the SolidWorks feature manager. This action causes the pop-up menu shown in Fig.11 to open.

3. Above this menu select the **Unsuppress** icon at arrow in Fig. 11. Immediately ¼ of the model is displayed in the graphics area.

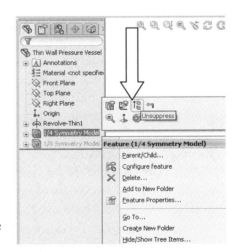

Figure 11- **Unsupressing** the symmetrical model in the SolidWorks feature manager.

4-11

4. Repeat steps 2 and 3, but this time select the grayed-out **1/8 Symmetry Model** and **Unsuppress** it.

The portion of the model selected for analysis should appear as shown in Fig. 12. Note that virtually any fraction of the model could be selected, but the 1/8 model is useful for reasons described later.

5. Toggle back to the **COSMOSWorks** analysis manager (i.e., return to COSMOSWorks by selecting the COSMOSWorks icon).

Notice how the use of descriptive names in the SolidWorks feature manager facilitated identification and selection of desired model attributes in the preceding steps.

The next step is to define restraints and loads applied to the model as outlined below.

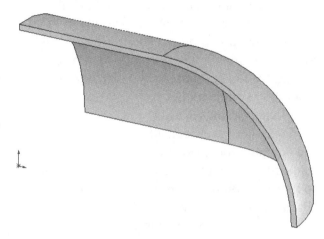

Figure 12 – A 1/8 model of the thin-wall pressure vessel. Symmetry is used to reduce analysis time and computer memory requirements.

6. Right-click the **Load/Restraint** icon and from the pull-down menu select **Restraints...**. The **Restraint** property manager opens.

7. In the **Type** dialogue box, click to open the pull-down menu and from the list of available boundary conditions, select **Symmetry**.

Just below the **Symmetry** boundary condition, the **Planar Faces for Restraint** field is highlighted (light blue) to indicate it is active and awaiting user selection of faces to which symmetry restraints are to be applied.

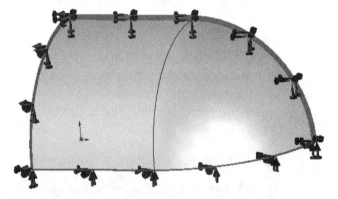

Figure 13 – An internal view of the model showing **Symmetry** restraints applied to all cut faces.

Pressure Vessels

8. Rotate and zoom-in on the model to select all *cut faces* shown in Fig. 13. Do *not* select edges or vertices on the model. The three selected faces, **Face<1>**, **Face<2>** and **Face<3>** appear in the **Planar Faces for Restraint** field.

Carefully observe the symbols used to indicate symmetry boundary conditions. All previous examples used solid tetrahedral elements for which **Immovable** restraints were appropriate. **Immovable** restraints prevent translations in the three coordinate directions X, Y and Z and are pictured in Fig. 14 (a). Symbols of this type are observed acting *normal* to all cut surfaces on the model. When fully restrained, shell elements restrict three translations in X, Y and Z directions *plus* they prevent rotation about the X, Y and Z axes. **Fixed** restraints are illustrated in Fig. 14 (b) where the added "disk" on the tail of each vector represents an added rotational restraint. Finally, the **Symmetry** restraint is pictured in Fig. 14 (c). It prevents translation in a direction normal to the restrained face (arrow shaped vector) *and* prevents rotation about the other two axes associated with the restrained face ("thumb-tack" appearance vectors). Examination of restraint symbols on the three cut faces reveals that all symbols pictured below are applied, however, each has a unique outer normal and rotational vector set corresponding to the surface on which they act.

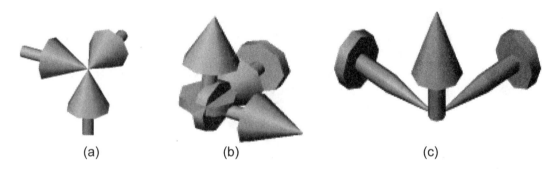

(a) (b) (c)

Figure 14 – (a) **Immovable** restraint symbol; (b) **Fixed** restraint symbol only applies to shell elements, however it need not always be used; (c) **Symmetry** restraint applied to shell elements on cut surfaces of the thin-wall pressure vessel.

9. Click **[OK]** ✓ to close the **Restraint** property manager.

Pressure Load Applied
Loading of the thin-wall cylinder is completed by the addition of an internal pressure as outlined next.

1. Right-click the **Load/Restraint** icon and from the pull-down menu select **Pressure…**. The **Pressure** property manager opens.

2. In the **Type** dialogue box, choose ⦿**Normal to selected face** (if not already selected).

3. The **Faces for Pressure** field is highlighted (light blue) to indicate it is active. Proceed to select the *inside* surfaces of both the cylindrical and hemispherical

portions of the model. **Face<1>** and **Face<2>** should appear in the **Faces for Pressure** field and both surfaces appear highlighted to indicate their selection.

4. Within the **Pressure Value** dialogue box, verify that units are set to **N/m^2** and beneath this field, in the **Pressure value** field, type **1.4e6** for the applied internal pressure. Units of **N/m^2** appear adjacent to this field.

Arrows should appear along all edges of the model as shown in Fig. 15. If the pressure arrows are not directed *toward the inner surface*, check **Reverse direction**.

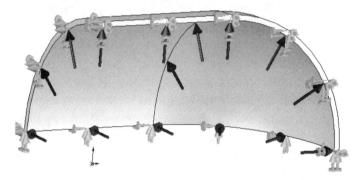

Figure 15 – Uniform pressure distribution shown acting on internal surfaces of the model.

5. Click **[OK]** ✓ to close the **Pressure** property manager.

Because the 1/8 model was selected, symmetry restraints restrict model displacement in the X, Y and Z directions and rotations about the same three axes. Therefore, no additional restraints need be applied. However, if the model were not fully restrained, a warning would appear during the solution and additional restraints to prevent "rigid body motion" would need to be applied. The need for rigid body restraint is demonstrated in the next example.

Meshing the Model

Recall that mesh quality (high) and the type of finite element solver (shell mesh) were pre-selected at the beginning of this example. Therefore, the meshing process outlined below is considerably shortened.

1. In the COSMOSWorks manager, right-click the **Mesh** folder and from the pull-down menu select **Create Mesh…**. The **Mesh** property manager opens.

2. Accept the default mesh size and click **[OK]** ✓ to close the **Mesh** property manager.

Meshing proceeds automatically and Fig. 16 shows an image of the thin-wall pressure vessel displayed with a *shell* mesh and all applied loads and restraints. Notice that the model now appears as a "paper-thin" shell.

Consistent with system default conventions adopted at the beginning of this example, the color orange represents the "bottom" surface of the shell elements. If the inner surface of the model is considered as the "bottom," then surfaces appearing orange on your model may have to be "flipped" (i.e., reversed). Do this only if necessary using the following steps.

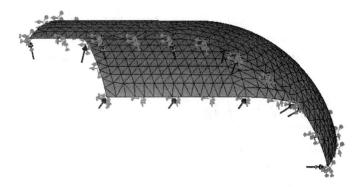

Figure 16 – A **Shell** mesh applied to the mid-surface of the pressure vessel model. Note that "top" and "bottom" surface designations are reversed.

3. Right-click the **Mesh** icon and from the pull-down menu select **Flip Shell Elements**. A COSMOSWorks warning window appears and prompts, **Select a face**. Click **[OK]** and proceed as follows.

4. While pressing the **[Shift]** key, click both the cylindrical *and* hemispherical surfaces on either the inside or outside of the model. Two information "flags" appear along with a vector normal to the last selected surface. Because these are only used to change the appearance of selected surfaces, click ⊠ to close them.

5. Then, once again right-click the **Mesh** icon and from the pull-down menu select **Flip Shell Elements**. *NOTE: It may be necessary to repeat step 5 to cause the desired changes to take effect*

6. Move the cursor well away from the model on the graphics screen and click to erase any remaining normal vectors or coordinate systems displayed.

Colors of the top and bottom surfaces should now reverse, (orange moves to the inside [bottom] and gray is switched to the outside [top] surface). The last three steps are not mandatory. They simply illustrate how the model can be altered to conform to the users' preference. The model is now complete and ready to be solved as outlined below.

Solution

1. Right-click **Pressure Vessel-Shell Study (-Default-)** and from the pull-down menu select **Run**. After the solution is complete, the three system default plots, **Stress1 (-vonMises-)**, **Displacement1 (-Res disp-)**, and **Strain1 (-Equivalent-)** are listed at the bottom of the COSMOSWorks manager.

Before continuing, toggle back to SolidWorks and observe that a new icon labeled **MidSurface1** appears at the bottom of the SolidWorks feature manager tree. If the "+" sign adjacent to **MidSurface1** is selected, two folders labeled **Surface-Imported1**, and **Surface-Imported2** appear. Next, move the cursor over each of these names and observe corresponding portions of the model highlighted on the graphics screen.

Analysis of Machine Elements using COSMOSWorks

These two mid-surfaces correspond to new geometries defined because **Shell mesh using mid-surfaces** was selected as the Study type at the beginning of this example. Toggle back to the COSMOSWorks manager and continue below.

Results Analysis

Because the emphasis of this example is on mastering basic techniques for working with shell elements, the review of final results is rather brief. Thus, a quick look at results of this analysis should confirm the accuracy of this approach. Only stress results are examined below.

1. Right-click the **Results** folder and from the pull-down menu select **Define Stress Plot…**. The **Stress Plot** property manager opens.

2. In the **Display** dialogue box, click to open the **Component** pull-down menu and from the list of stresses, select **P1: 1st Principal Stress.**

3. Click **[OK]** ✓ to close the **Stress Plot** property manager and a plot of 1st principal stress, like that shown in Fig. 17, appears in the graphics area. Also, **Stress2 (-1st principal-)** is added beneath the **Results** folder.

For convenience in viewing results, turn off the load and restraint symbols. Try this on your own. If you encounter difficulty, proceed as follows.

4. Right-click the **Load/Restraint** icon and from the pull-down menu select **Hide All**.

Figure 17 – First principal stress distribution over a 1/8 model of the thin-wall pressure vessel. Note the stress gradient in the transition region between the cylinder and hemispherical end.

5. Return to the **Stress2 (-1st principal-)** folder and double-click it. The plot should now appear similar to that illustrated in Fig. 17.

Taking the maximum value for $\sigma_1 = \sigma_t = 3.429e+007$ N/m^2 from the **Stress2 (-1st principal-)** plot and comparing it with the value calculated using equation [1] for stress in a thin wall pressure vessel $\sigma_t = 3.36e7$ N/m^2, the following percent difference is determined.

$$\% \text{ difference} = \left[\frac{\text{FEA result} - \text{classical result}}{\text{FEA result}}\right] 100 = \left[\frac{3.429e7 - 3.36e7}{3.429e7}\right] 100 = 2.01\% \quad [3]$$

Next, recalling that tangential stress in the hemispherical heads is ½ σ_t in the cylindrical portion, the following comparison is made for stresses in the hemispherical head. The value of stress for this comparison is determined by selecting an arbitrary number of points on the hemispherical head using the **Probe** tool as shown in Fig. 18. Try this on your own. The following steps provide guidance if desired.

6. Right-click **Stress2 (-1st principal-)** and from the pull-down menu, select **Probe**. The **Probe Result** property manager opens as shown in Fig. 18.

7. In the **Options** dialogue box select ⊙ **On selected entities**.

8. In the **Results** dialogue box for **Faces, Edges or Vertices** is highlighted (light blue). Move the cursor onto the hemispherical end of the model and click to select it. **Face<1>** appears in the **Faces, Edges or Vertices** field.

9. Near the middle of the **Results** dialogue box, click the **[Update]** button and immediately the table is populated with stress magnitudes at all nodes on the selected surface.

10. In the **Summary** dialogue box, near bottom of the property manager, observe the average (**Avg**) value of stress on this surface is 1.6408e+007 N/m^2. Be aware that this value includes a portion of the higher stresses near the hemispherical to cylindrical transition region on the model.

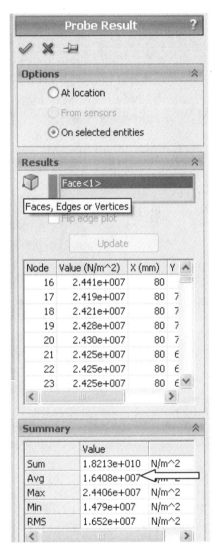

Figure 18 – The **Probe Results** table showing stress magnitudes on the hemispherical end of the pressure vessel.

The value of stress determined using classical stress equations is determined by taking half of the value of $(\sigma_t)_{\text{hemispherical}}$ = ½ * $(\sigma_t)_{\text{cylinder}}$ = ½ * 3.36e7 Pa = 1.68e7 N/m². The percent difference between finite element and classical results yields the following comparison for stress magnitude in the hemispherical end.

$$\% \text{ difference} = \left[\frac{\text{FEA result} - \text{classical result}}{\text{FEA result}}\right]100 = \left[\frac{1.6408e7 - 1.68e7}{1.6408e7}\right]100 = 2.4\% \quad [4]$$

An alternate way of determining stress in the hemispherical end (or anywhere on any model) is described next.

11. Return to the **Options** dialogue box and select ⊙ **At location**. The format of the **Probe Result** property manager changes and all values are cleared from the tables.

12. Move the cursor onto the hemispherical end of the model and click to select several nodes at points in the dark blue, color-coded area as shown in Fig. 19. Select points well away from the junction between the hemispherical and cylindrical surfaces.

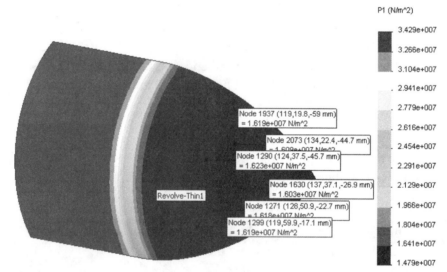

Figure 19 – Randomly selected nodes used determine an average value of stress on the outer surface of the hemispherical end.

Manually compute the average stress based on values listed in the **Results** dialogue box. The percent difference between Finite Element and classical equations is shown in equation [5]. Not surprisingly the percent difference is greater because higher stress magnitudes near the transition region are excluded from this calculation. (Results will vary.)

$$\% \text{ difference} = \left[\frac{\text{FEA result} - \text{classical result}}{\text{FEA result}}\right]100 = \left[\frac{1.62e7 - 1.68e7}{1.62e7}\right]100 = 3.7\% \quad [5]$$

13. Click **[OK]** ✓ to close the **Probe Result** property manager.

Pressure Vessels

Next, display the value of second principal stress and compare it against the value computed for longitudinal stress calculated using classical equations; see equation [2] on page 4-2. Try creating a plot of second principal stress on your own. If guidance is needed, proceed as follows.

14. Right-click the **Results** folder and from the pull-down menu select **Define Stress Plot…**. The **Stress Plot** property manager opens.

15. In the **Display** dialogue box, click to open the **Component** pull-down menu and from the list of stresses, select **P2: 2nd Principal Stress.**

16. Click **[OK]** ✓ to close the **Stress Plot** property manager and a plot of 2nd principal stress appears in the graphics area.

The green color-coded stress magnitude at both the mid-section of the cylinder and the hemispherical end corresponds to the second principal stress. Based on the color chart, its value is approximately[2] $\sigma_2 = \sigma_\ell = 1.62e7$ N/m^2. As expected, this value of longitudinal stress is equal to half of the tangential stress. Therefore, further exploration of this stress is not pursued here.

Notice that σ_1 and σ_2 observed above occur on the *top* (i.e., outside) surface of the model. To illustrate that a slightly different state of stress occurs on the *bottom* (i.e., inner) surface of the model, return to **Stress2 (-1st principal-)** to investigate this point.

17. Double-click **Stress2 (-1st principal-)** to again display the plot of σ_1.

The maximum and minimum values of σ_1 displayed on the current plot are for stress on the *top* (outside) surface of the model. Those values should be approximately : $(\sigma_1)_{Max} = 3.429e+007$ N/m^2 and $(\sigma_1)_{Min} = 1.479e+007$ N/m^2. Next, examine stresses on the bottom (inside) surface.

18. Right-click **Stress2 (-1st principal-)** and from the pull-down menu select **Edit Definition…**. The **Stress Plot** property manager opens as shown in Fig. 20.

19. At the bottom of the **Display** dialogue box, the **Shell Face** field currently selected for display is the **Top** surface. Change this, in the pull-down menu, by selecting **Bottom**.

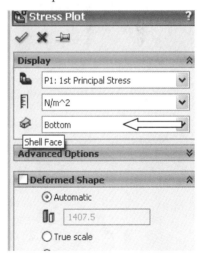

Figure 20 – Changing selection of the **Shell Face** viewed on the model.

---

[2] The magnitude $\sigma_2 = \sigma_\ell$ was determined by averaging the stress magnitudes at top and bottom of the matching green portion of the color coded stress legend $\sigma_2 = \sigma_\ell = (1.577e7 + 1.662e7)/2 = 1.62e7$ N/m^2.

4-19

20. Click **[OK]** ✓ to close the **Stress Plot** property manager.

> **Analysis Insight**
>
> Stress magnitudes corresponding to σ_1 on the bottom (i.e., inner surface) of the pressure vessel now show the following values: $(\sigma_1)_{Max} = 3.603\text{e}+007$ N/m^2 and $(\sigma_1)_{Min} = 1.795\text{e}+007$ N/m^2. Based on different results occurring on these two surfaces, two observations are made.
>
> - First, stress on the inside surface of the pressure vessel is greater than that observed on the outer surface.
>
> - Second, unlike stresses on different surfaces of a solid model that can be observed by rotating the model, stresses on top and bottom surfaces of shell elements can only be observed by altering the **Stress Plot** definition of the surface observed.

At this point the *shell* element example is concluded. Before analyzing the thick-wall pressure vessel, select **File / Close**. When prompted to "**Save changes to Thin Wall Pressure Vessel**," select **[No]** to exit the thin-wall pressure vessel example without saving results.

THICK WALL PRESSURE VESSEL

In this second-half of the current chapter, analysis of a thick-wall pressure vessel is performed. Thick-wall pressure vessels are found in numerous applications including high pressure piping, gun and cannon barrels, hydraulic cylinders, and related uses. Thick-wall pressure vessels are classified as such in applications where wall thickness "t" exceeds $t > r_i/20$, where r_i = inside radius. For these applications, assumptions related to thin-wall pressure vessels are no longer valid. Primary among these differences are: (a) radial stress, which is neglected in thin-wall formulations, is known to vary through the thickness of a thick-wall vessel; and (b) tangential (a.k.a., hoop or circumferential) stress also varies through the wall thickness. Much of thick-wall cylinder theory is also relevant to analysis of press and shrink fits, which are investigated in the next chapter.

Problem Statement

Thick-Wall Pressure Vessel

This example is based on a thick-wall cylindrical pressure vessel that is closed on one end, but attached to a rigid pipe connection at the opposite end as illustrated in Fig. 21. This vessel is made of Alloy Steel and is subject to an internal pressure $P_i = 16.0$ MPa and an external pressure that approximates standard atmospheric pressure $P_o = 0.101$ MPa. Cylinder dimensions, necessary to apply classical stress equations, are also included in Fig. 21. The goal is to determine tangential and radial stress variation through the cylindrical walls of the pressure vessel.

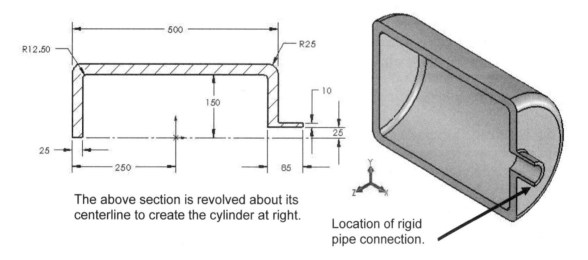

Figure 21 – Basic dimensions and geometry of the thick-wall pressure vessel.

1. Open SolidWorks by making the following selections. (*Note:* "/" is used to separate successive menu selections.) *Skip to step 2 if SolidWorks is already open.*

   Start / All Programs / SolidWorks 2008

2. From the main menu within SolidWorks select **File / Open**. Then use procedures common to your computer environment to open the COSMOSWorks file named **Thick Wall Pressure Vessel**.

Defining the Study

Because tetrahedral elements are used to create a solid model of the pressure vessel, the majority of procedural steps employed to create this study are familiar to users who have worked the preceding examples. However, all necessary steps are included. This study again uses symmetry restraints and also introduces new methods of viewing results.

1. Click to select the **COSMOSWorks** manager icon.

2. Right-click **Thick Wall Pressure Vessel** and from the pull-down menu select **Study...** The **Study** property manager opens.

3. In the **Name** field, type: **Thick Wall Pressure Vessel**.

4. Select **Solid mesh** (if not already selected) and in the **Type** dialogue box ensure that a **Static** study is selected.

5. Click **[OK]** ✓ to close the **Study** property manager. An outline of study components appears beneath the COSMOSWorks manager.

Unlike the 1/8 model used for the thin-wall pressure vessel, a 1/2 model is selected for the thick-wall example. Notice also that complete geometric symmetry does not exist for this model (left and right ends of the model differ). Return to **SolidWorks** where the model is defeatured and where half of the model is suppressed. For purposes of illustration, the small fillet at the pipe connection is selected to be defeatured. Proceed as follows.

6. Toggle to the **SolidWorks** feature manager design tree.

7. At the bottom of the SolidWorks feature manager, right-click **Pipe Connector Fillet** and select the **Suppress** icon. The fillet is removed at the junction between the cylinder and pipe extension as illustrated in Fig. 22.

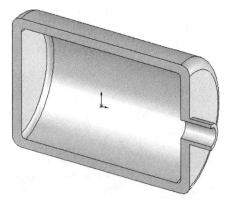

Figure 22 – Half-section of the thick-wall pressure vessel after defeaturing.

Pressure Vessels

8. Near the bottom of the **SolidWorks** feature manager, right-click **Half-Model of Cylinder** and from the pop-up menu, select the **Unsuppress** icon.

The model image changes to the half-cylinder shown in Fig. 22. Notice how assigning descriptive names to folders facilitates identifying their function. Before renaming the **Half-Model of Cylinder** folder, its generic name was "Cut-Extrude2."

9. Toggle back to the **COSMOSWorks** analysis manager where the remainder of this study is defined.

Assign Material Properties

1. Right-click the **Solids** folder and from the pull-down menu, select **Apply Material to All...**. The **Material** window opens.

2. Because users should be familiar with this window, on your own select ⊙ **From library files** and choose **Alloy Steel (SS)**. Verify that units are set to **SI**, then click **[OK]** to close the window. A check mark "✓" appears of the **Solids** folder.

Define Restraints and Loads

1. Right-click **Load/Restraint** and from the pull-down menu choose **Restraints...**. The **Restraint** property manager opens.

2. In the **Type** dialog box, open the pull-down menu and select **Symmetry**. Next, select the cut surface highlighted in Fig. 23.

3. Click **[OK]** ✓ to close the **Restraint** property manager.

Restraint symbols in the Z-direction appear on the model. Physically this means that existing symmetry restraints permit axial displacements (along the length of the cylinder) and radial displacements (change of cylinder diameter) when the cylinder is subject to internal or external pressure. These displacements are consistent with the half-model selected. At present, however, both X and Y displacements of the model are still possible because the model is not fully restrained.

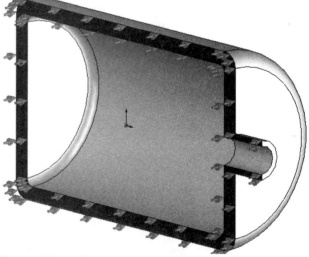

Figure 23 – Half-model of pressure vessel with symmetry restraints applied in the Z-direction.

4-23

Analysis of Machine Elements using COSMOSWorks

Because symmetry restraints restrain the model in the Z-direction only, it is necessary to apply additional restraints in the X and Y directions to prevent what is known as "rigid body" motion. The type of restraints selected depends heavily upon the nature of actual operating conditions. Because the right end of the pipe connection is considered attached to a rigid (i.e., immovable) pipe extension (not shown) the remaining restraints are applied at that location. Fortunately, this immovable restraint does not interfere with finite element results in the thick-wall portion of the pressure vessel.

Analysis Insight
Although not part of this example, consider the following two scenarios:

A different set of restraints results if, in addition to being attached to a rigid pipe at its right end, the left end of the cylinder is mounted against a fixed surface (e.g., a wall or support bracket). That additional restraint would alter stress distribution within the cylinder because longitudinal deformation of the model is restrained from both ends.

Consider yet one additional design/analysis scenario in which the model, restrained as described in the preceding paragraph, is subsequently subjected to a significant temperature change. This condition would add temperature induced deformation (expansion or contraction) associated with the coefficient of thermal expansion for the model material. This scenario would require a separate analysis to determine thermally induced deformations.

Continue to define remaining restraints and loads in the following steps.

4. Right click the **Load/Restraint** folder and from the pull-down menu select **Restraints…**. The **Restraint** property manager opens.

5. Beneath **Type**, select **Immovable (No translation)**.

6. Zoom in on the right-end of the pipe extension, shown in Fig. 24, and click to select this surface. Immovable restraint symbols in the X, Y and Z directions are added to the model and **Face<1>** appears in the **Type** dialogue box.

7. Click **[OK]** ✓ to close the **Restraint** property manager.

    The model is now fully restrained in all directions.

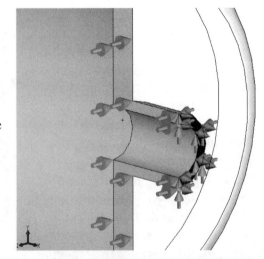

Figure 24 – The **Immovable** restraint is added to the attachment surface associated with a rigid pipe connection.

The next step is to apply an internal pressure $P_i = 16.0$ MPa to all internal surfaces of the pressure vessel. Proceed as follows.

8. Right-click **Load/Restraint** and from the pull-down menu select **Pressure…** The **Pressure** property manager opens.

9. In the **Type** dialogue box, select ⊙**Normal to selected face**.

10. The **Faces for Pressure** field is highlighted (light blue) to indicate it is active and awaiting user input.

11. Move cursor onto the model and click to select *all interior* surfaces (both cylinder ends, both internal fillets, and cylindrical surfaces of the cylinder and the pipe extension). A total of six faces (**Face<1>, Face<2>, … Face<6>**) should be listed in the **Faces for Pressure** field.

12. Beneath **Pressure Value**, verify that **Units** are set to **N/m^2** and type **16.0e6** in the **Pressure Value** field.

Figure 25 – Internal pressure applied to six faces within the thick-wall pressure vessel.

13. Click **[OK]** ✓ to close the **Pressure** property manager. The model should appear as shown in Fig. 25.

Next apply atmospheric pressure $P_o = 0.101$ MPa to all external surfaces of the model. Try this on your own or follow steps below.

14. Repeat steps 8 through 13 with the following two exceptions.

    a. In step 11, select *all exterior* surfaces. Once again, a total of six faces should be chosen.

    b. In step 12, type **0.101e6** N/m^2 in the **Pressure Value** field.

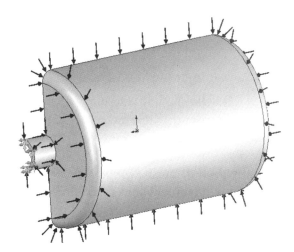

Figure 26 – Atmospheric pressure applied to exterior of the thick-wall pressure vessel.

Upon completion of step 14 (a) and (b), a view of the exterior of the pressure vessel should appear as shown in Fig. 26.

Mesh the Model

1. Right-click the **Mesh** folder and from the pull-down menu select **Create Mesh...** The **Mesh** property manager opens.

2. At top of **Mesh Parameters** dialogue box accept the default mesh size (slider pointer located at middle of the **Coarse/Fine** indicator scale) and verify that **Units** are expressed in **mm**. If not, change them using the pull-down menu.

3. At bottom of the property manager, click [▽] to open the **Options** dialogue box.

4. Within this dialogue box verify that default settings appear as follows.

   a. Mesh **Quality:** should be set at ⦿**High**.

   b. **Mesher:**, should be set at ⦿ **Standard**.

   c. **Mesher options:** check ☑ **Jacobian check for solid** and set it to **4 points**.

5. Click **[OK]** ✓ to close the **Mesh** property manager.

The meshed model along with its boundary conditions is shown in Fig. 27.

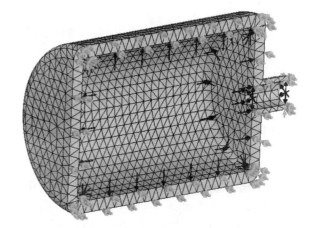

Figure 27 – Default mesh size applied to the thick-wall pressure vessel.

Analysis Insight

Examination of mesh size on the *cut surface* of the pressure vessel in Fig. 27 reveals that a single element spans the entire wall thickness. However, near the beginning of this chapter, a general guideline for determining when *solid* elements can be used stated that, "...the minimum number of elements recommended across the thickness of a part (i.e., across the part's minimum dimension) should be *at least two elements.*" Given this guideline, it is appropriate to reduce element size so that at least two elements span the thickness of the pressure vessel wall. This change is easily made by returning to the **Mesh** property manager and adjusting mesh size as outlined below.

Pressure Vessels

6. Right-click the **Mesh** folder and from the pull-down menu select **Create Mesh…**. The **Mesh** property manager opens

7. In the **Mesh Parameters** dialogue box click-and-*drag* the pointer on the **Mesh Factor** sliding-scale from its default, mid-range position, to the **Fine** position.

8. Click **[OK]** ✓ to close the **Mesh** property manager and initiate re-meshing the model. Note the increased time required to mesh the model. The model now appears as shown in Fig. 28 with two elements across the wall thickness.

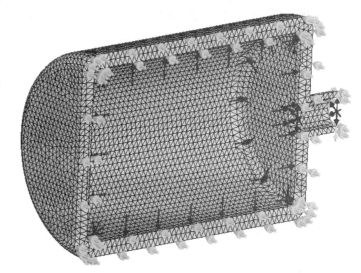

Figure 28 – Pressure vessel showing mesh size reduced to half of the default mesh size so at least two elements span the cylinder wall thickness.

Analysis Insight

Based on the stress concentration example of Chapter 3, where variation of mesh size was investigated, it was found that moving the pointer on the mesh size slider-scale to either the left or right extreme positions had the effect of doubling or halving the default mesh size.

The default mesh size for the model shown in Fig. 27 is **20.060353 mm**. Thus, changing element size to **Fine** reduces its nominal size to **10.030176 mm**, or exactly half the previous size. As a consequence, two elements now span the thick-wall portion of the pressure vessel as illustrated in Fig. 28.

Although two elements span the thick-wall portion of the model, only one element spans the thickness in the pipe connector neck. Zoom in and examine the neck area. This outcome might be considered acceptable because tangential and radial stresses in the thick-wall portion of the pressure vessel are of primary importance in this Study.

When compared to the original model, however, a dramatic increase in the number of nodes and elements is observed for the re-meshed model. These changes are summarized in Table 1.

Aside:
The default mesh size would yield adequate results for this model. However, because stress through the wall thickness is examined, the additional nodes will prove helpful.

Table 1 – Comparison of Finite Element Model Size

	Original Model (Default Mesh Size)	Revised Model (½ Mesh Size)	Increase in Model Complexity
No. of Nodes	16,085	88,194	Approx. 5.5 times larger
No. of Elements	9,364	55,978	Approx. 6 times larger

In this instance, following the general guideline, which suggests at least two solid elements across a part's minimum thickness, results in a significant increase in mesh density with a corresponding increase in solution time. This is a logical cause-and-effect consequence of a mesh size change.

Solution

Having defined restraints, loads, and a mesh, the thick-wall model is subjected to analysis as outlined below.

1. Right-click the study folder **Thick Wall Pressure Vessel (-Default-)** and from the pull-down menu, select **Run**. Solution time is significantly longer, approximately one minute on a Pentium 4® personal computer.

2. Computed results, in the form of plots, are listed beneath the **Results** folder. Notice that plots contained in these folders revert to the system default plots.

Results Analysis

Displacement Analysis

To gain some insight into the deformation of the thick-wall pressure vessel subject to internal and external pressures, attention is first directed to the **Displacement1 (-Res disp-)** folder.

1. Right-click the **Displacement1 (-Res disp-)** folder and from the pull-down menu select **Show**. A displacement plot of the model appears as seen in Fig. 29.

2. From the COSMOSWorks manager tree, right-click **Load/Restraint**, and from the pull-down menu select **Hide All**. This action removes all loads and restraints from the display.

3. Unfortunately, the previous step also removes the color-coded image of displacement in the pressure vessel. Therefore, either double-click **Displacement1 (-Res disp-)** or right-click it and select **Show** from the pull-down menu. Fig. 29 is again displayed. *Orient the model in a front view.*

4. Move the cursor over the screen image to display an *un-deformed* outline of the model as super-imposed on Fig. 29. Placing the cursor near the mid-point of a cylindrical wall works well.

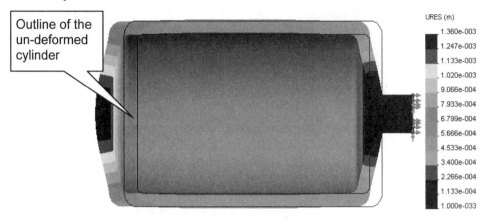

Figure 29 – Image of the model with symmetry and pressure boundary conditions "hidden." **Immovable** restraints and the un-deformed shape of the model are superimposed on the image.

A comparison of the displacement plot with the un-deformed shape of the pressure vessel reveals the following, common sense, observations.

a. All displacements are away from the immovable pipe-end.

b. Bulging of both the left and right ends of the model contributes to longitudinal (axial) deformation away from the immovable end.

c. Longitudinal deformation (axial stretch) of the cylindrical section also contributes to overall axial deformation.

d. Slight radial deformation (bulging) of cylindrical walls is observed.

e. Virtually zero displacement occurs in the pipe connection.

Analysis Insight

Taking into account the fact that displacements depicted in Fig. 29 are greatly exaggerated, briefly consider how the different boundary conditions (such as supports or a fixed wall at the left-end of the pressure vessel) described in an earlier **Analysis Insight** section, would impact results of the current example.

The effect of correct or incorrect boundary conditions (i.e., loads and restraints) on finite element analysis results cannot be overemphasized!

Analysis of Machine Elements using COSMOSWorks

Von Mises Stress Analysis

The next topic for investigation is interpretation of stresses occurring within the model.

1. Double-click the **Stress1 (-vonMises-)** folder to display the plot shown in Fig. 30.

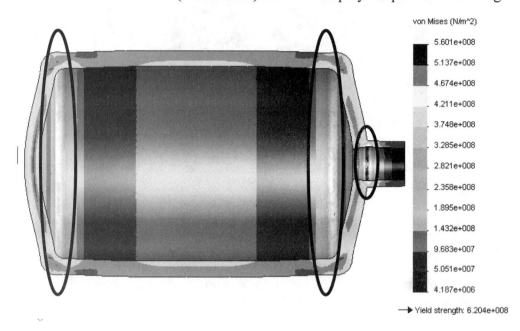

Figure 30 – Regions of high von Mises stress on the thick-wall pressure vessel model.

Regions of high von Mises stress at fillet radii on both ends of the cylinder and in the vicinity of the pipe-to-cylinder connection are circled on Fig. 30. All von Mises stresses are below the material yield strength, which is listed beneath the color coded stress chart.

Tangential Stress Analysis

The next focus of this study is on tangential and radial stress distributions typically associated with analysis of pressurized thick-wall cylinders. However, before examining stress plots, it is helpful to have some expectation of magnitudes associated with these two stresses. Therefore, solutions based on Lame's equations for values of tangential and radial stresses at both the inside and outside surfaces of a thick-wall cylinder are included below.

Tangential stress at outside surface:

$$(\sigma_t)_o = \frac{p_i r_i^2 - p_o r_o^2 - r_i^2 r_o^2 (p_o - p_i)/r_o^2}{r_o^2 - r_i^2}$$

$$= \frac{16.0e6(0.15)^2 - 0.101e6(0.175)^2 - (0.15)^2(0.101e6 - 16.0e6)}{(0.175)^2 - (0.15)^2} = 87.9 \text{ Mpa}$$

[6]

Tangential stress at inside surface:

$$(\sigma_t)_i = \frac{p_i r_i^2 - p_o r_o^2 - r_i^2 r_0^2 (p_o - p_i)/r_i^2}{r_o^2 - r_i^2}$$

$$= \frac{16.0e6(0.15)^2 - 0.101e6(0.175)^2 - (0.175)^2(0.101e6 - 16.0e6)}{(0.175)^2 - (0.15)^2} = 103.8 \text{ Mpa}$$

[7]

Radial stress at inside surface:

$$\sigma_{r_i} = \frac{p_i r_i^2 - p_o r_o^2 + r_i^2 r_0^2 (p_o - p_i)/r_i^2}{r_o^2 - r_i^2} = \frac{p_i r_i^2 - \cancel{p_o r_o^2} + \cancel{p_o r_o^2} - p_i r_o^2}{r_o^2 - r_i^2}$$

$$= \frac{-p_i (r_o^2 - r_i^2)}{(r_o^2 - r_i^2)} = -p_i = -16.0 \text{ Mpa} = \text{(internal pressure)}$$

[8]

Radial stress at outside surface:

$$\sigma_{r_o} = \frac{p_i r_i^2 - p_o r_o^2 + r_i^2 r_0^2 (p_o - p_i)/r_o^2}{r_o^2 - r_i^2} = \frac{\cancel{p_i r_i^2} - p_o r_o^2 + p_o r_i^2 - \cancel{p_i r_i^2}}{r_o^2 - r_i^2}$$

$$= \frac{-p_o (r_o^2 - r_i^2)}{(r_o^2 - r_i^2)} = -p_o = -0.101 \text{ MPa} = \text{(external pressure)}$$

[9]

Two alternative ways of viewing these results are investigated below. We begin with what might be considered the most intuitive method of viewing stress results and then proceed to a second method that makes use of the *Section Clipping* capability of the software. Begin by adding a plot of first principal stress to the **Results** folder. Try this on your own or follow the steps outlined below.

1. Right-click the **Results** folder and from the pull-down menu, select **Define Stress Plot…** The **Stress Plot** property manager opens.

2. In the **Display** dialogue box, click to open the **Component** pull-down menu and from the list of available stresses, select **P1: 1st Principal Stress**.

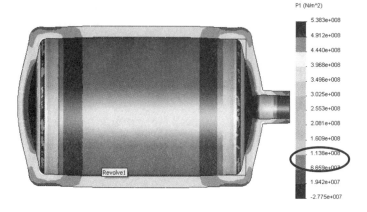

Figure 31 – Plot of 1st principal stress distribution throughout the thick-wall pressure vessel.

3. Click **[OK]** ✓ to close the **Stress Plot** property manager.

Figure 31 shows a front view of the distribution of 1st principal stress throughout the model. Once again notice regions of high stress at both end fillets where the cylinder and flat ends join. Unfortunately, Fig. 31 does not provide good delineation of stress variation through the central, thick-wall portion of the cylinder.

Analysis Insight

Regions of high principal stress and maximum von Mises stress are typically most important to the designer or stress analyst whose focus is on product safety. However, because a primary objective of this example is to investigate stresses associated with thick-wall pressure vessel theory and to learn different ways to display these results, the following section describes how to better examine these stresses while simultaneously discovering additional software plotting capabilities.

Adjusting Stress Magnitude Display Parameters

A quick review of previous calculations for tangential stress variation from the inside to the outside cylindrical surfaces of the model, equations [6] and [7] above, reveals:

Tangential stress at inside: $(\sigma_t)_i = 103.8$ Mpa

Tangential stress at outside: $(\sigma_t)_o = 87.9$ MPa

Comparing the above values with the range of 1st principal stress magnitudes displayed in the color-coded stress legend adjacent to the model (currently displayed on your screen) reveals that the above stresses occupy only a small portion of the full range of stress values displayed; see circled portion of the stress legend on Fig. 31. Thus, if it is desired to verify values of 1st principal stress between 87.9 MPa to 103.8 MPa, in the cylinder wall, then the stress display must be adjusted to bracket these stress values. The following steps outline a procedure to accomplish this.

1. Right-click **Stress2 (-1st principal-)** and from the pull-down menu select **Chart Options…**. The **Chart Options** property manager opens as shown in Fig. 32.

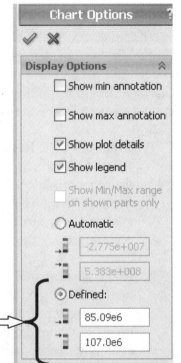

Figure 32 – Controlling the range of stresses displayed using the **Defined** option.

Pressure Vessels

2. At the bottom of the **Display Options** dialogue box, select ⦿**Defined:** This option provides a means of setting the desired upper and lower bounds for stress magnitudes to be displayed. In the upper box, type **85.0e6**, which is slightly lower than the *minimum* 1st principal stress value of $(\sigma_t)_o = 87.9$ MPa expected at the outside surface.

3. Similarly, in the bottom box, type **107.0e6**, which is slightly higher than the *maximum* 1st principal stress value of $(\sigma_t)_i = 103.8$ MPa expected at the inside surface of the cylinder.

4. Click **[OK]** ✓ to close the **Chart Options** property manager. Figure 33 (a) displays 1st principal stress magnitudes in the defined range.

5. Next, zoom-in on the boxed region, shown in Fig. 33 (a), to obtain a close-up view of stress variation through the cylinder wall as illustrated by the multiple color fringes in Fig. 33 (b). Notice that this region is well removed from effects of end-conditions caused by fillets at both ends of the cylinder.

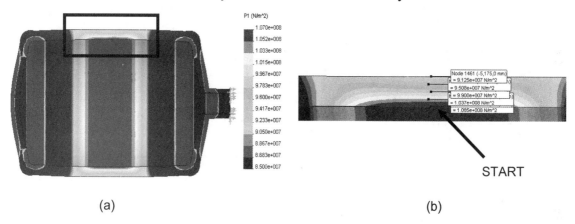

(a)   (b)

Figure 33 – (a) 1st principal stress plot after magnitudes are restricted to the range: 85.0 MPa < σ_t < 107.0 MPa; (b) close-up of tangential stress distribution (σ_t) through the thick-wall.

The **Probe** feature is used next to determine actual values of 1st principal stress at various nodes across the wall thickness.

6. Right-click **Stress2 (-1st principal-)** and from the pull-down menu select **Probe**. The **Probe** window opens as shown in Fig. 34 (a).

7. Begin at the inside of the cylinder wall, as near as possible to the middle of the cylinder; see "**START**" label on Fig. 33 (b). Then successively click and move the cursor slightly upward (moving in as straight a line as possible) and repeat until five node points are selected across the wall thickness. Stress values at each node point are simultaneously displayed on the model [also shown in Fig. 33 (b)] and in the **Probe** window, Fig. 34 (a). *If difficulty selecting nodes is experienced, see the ASIDE section below.*

> **ASIDE:**
> If difficulty is experienced selecting the *unseen* nodes, turn "on" the mesh display as follows.
>   a. Right-click **Stress2 (-1st principal-)** and from the pull-down menu select **Settings...**
>
>   b. Next, within the **Settings** property manager, click to open the pull-down menu beneath **Boundary Options** and select **Mesh**. Click **[OK]** ✓.
>
> A mesh displayed on the model should facilitate selecting corner and mid-side nodes on each element across the wall thickness. Repeat steps 6 and 7 with the mesh displayed.

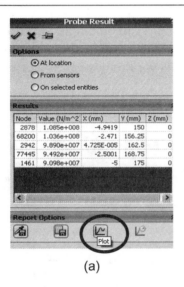
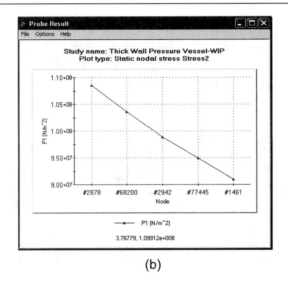

(a) (b)

Figure 34 – (a) **Probe** window displays values of $\sigma_1 = \sigma_t$ at node locations across the thick-wall; (b) the **Probe Result** window shows a graph of tangential stress variation from the inside surface (high stress) to the outside surface (lower stress) of the thick-wall pressure vessel.

8. After using the **Probe** feature to select nodes, click the **[Plot]** button at bottom of the **Probe** window to display a graph of tangential stress variation through the wall thickness as shown in Fig. 34 (b).

9. Click ☒ to close the **Probe Result** graph.

10. Click **[OK]** ✓ to close the **Probe Result** window.

11. If the mesh is still displayed on the model, right-click **Stress2 (-1st principal-)** and from the pull-down menu select **Settings**. In the **Boundary Options** dialogue box, open the pull-down menu and select **Model**, then click **[OK]** ✓.

A quick comparison of classical results, found using Lame's equations, and finite element results for tangential stress magnitudes at the inside and outside wall surfaces yields the following.

For tangential stress at the inside surface:

$$\% \text{ difference} = \left[\frac{\text{FEA result} - \text{classical result}}{\text{FEA result}}\right]100 = \left[\frac{108.5e6 - 103.8e6}{108.5e6}\right]100 = 4.33\% \quad [10]$$

At the outside wall surface:

$$\% \text{ difference} = \left[\frac{\text{FEA result} - \text{classical result}}{\text{FEA result}}\right]100 = \left[\frac{90.98e6 - 87.9e6}{90.98e6}\right]100 = 3.38\% \quad [11]$$

Both these values are deemed low enough that the analyst should have reasonable confidence in the finite element analysis results.

This concludes examination of tangential stress variation through the cylinder wall by conventional methods. The next section examines an alternate means of viewing these same results by using **Section** plots.

Using Section Clipping to Observe Stress Results

Use of *Section Clipping* is introduced in this section. Section plots are used where it is desirable to view stresses interior to a solid model, such as in an assembly of multiple parts, or if a solid model of the cylinder were used for this example rather than a cut-section of the model. Steps below outline use of the *Section Clipping* feature.

1. If the previous close-up view of the pressure vessel wall, appearing in Fig. 33 (b) remains on the screen, use the **Zoom to Fit** icon to restore the graphic display to a full image of the model. Then click the **Isometric** icon to reorient the model in an isometric view similar to that shown in Fig. 35.

2. Right-click **Stress2 (-1st principal-)** and from the pull-down menu select **Section Clipping…**. The **Section** property manager opens as shown in Fig. 36.

3. In the **Section1** dialogue box, notice that **Front Plane** is selected as the default viewing plane. Also, the **Distance** field, located immediately below **Front Plane** in Fig. 36, is set at **0.00 mm**, thereby indicating that the section view is displayed *on* the front surface.

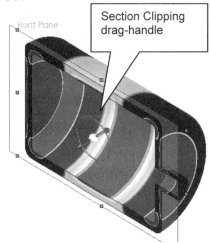

Figure 35 – Isometric view of model showing 1st principal stress and drag-handles for **Section Clipping** control.

4. In the **Options** dialogue box, clear the "✓" from ☐ **Show contour on the uncut portion of the model**. Immediately the interior and exterior surfaces of the model are shaded the default model color (typically gray). This makes for easier viewing of stress contours on cut sections.

5. Next, click-and-drag the *Section Clipping* drag-handles shown in Fig. 35. This action permits dynamic viewing of stresses within the model at any depth from the **Front Plane**.

6. Undo step 4 by checking ☑ **Show contour on the uncut portion of the model**, and again move the *Section Clipping* drag-handles.

A similar effect to that created in step 4 can be obtained as outlined next.

7. In the **Options** dialogue box, check to select ☑ **Plot on section only**. This action restricts plotted stresses to the **Front Plane** *at* the current depth. Once again, move the *Section Clipping* drag-handles and observe the resulting display.

8. To display stress distribution on a mid-plane through the model, type **0** in the **Distance** field of the **Section 1** dialogue box as shown in Fig. 36, and press **[Enter]**. This action creates the display shown in Fig. 37.

Figure 36 – The **Section** property manager is used to control display of stress contours *within* a model.

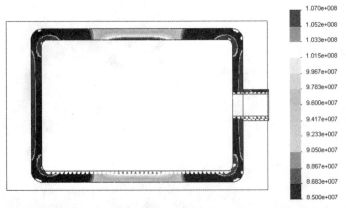

Section Clipping options demonstrated in steps 4 through 8 are convenient for masking stress contours which otherwise might make it difficult to differentiate between desired stress contours and other nearby colored fringes.

Figure 37 – Image of tangential stresses in the cylinder wall with the ☑ **Plot on section only** option active.

Pressure Vessels

9. Within the **Section1** dialogue box, experiment by clicking the down arrow ▼ adjacent to **0.00mm** in the **Distance** spin box. The down arrow moves the section away from the front plane in the negative z-direction. Click-and-*hold* the down arrow ▼ to observe the section plane move progressively away from the cut section and *into* the model. Conversely, if the up arrow ▲ is clicked, the section viewing plane moves away from the model (+ z-direction). For values greater than **0.00mm**, the section plane is located *in front* of the model, thus the model is not sectioned. It is also possible to examine stress at a particular depth within the model by typing any desired location into the **Distance** spin-box.

10. Reset **Distance** to **0.00** mm and clear the check-mark "✓" from ☐ **Plot on section only**.

As a final demonstration of the *Section Clipping* feature, tangential stresses through the cylinder wall are viewed by sectioning the cylinder as outlined below.

11. Either rotate the model to display its right-side view, or, from the SolidWorks toolbar, select the right-side view icon. A right-side view, looking down the axis of the cylinder from the pipe connection-end, appears in Fig. 38.

12. In the **Section 1** dialogue box, click to activate the **Reference entity** field (top field that currently shows **Front Plane**).

13. Next, click the "+" sign adjacent to the **SolidWorks** flyout menu, at top-left of the graphics screen, and from the pull-down menu, select **Right Plane**.

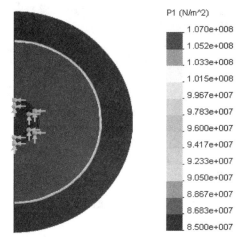

Figure 38 – Right-side view of model prior to taking a section view.

14. Notice that **Right Plane** now appears in the highlighted **Reference Entity** field and, because the origin of the Cartesian coordinate system is located at the middle of the model, the right plane bisects the model at its mid-section. As a result, tangential stress variation through the wall thickness is shown as illustrated in Fig. 39. Rotate the model to verify this view.

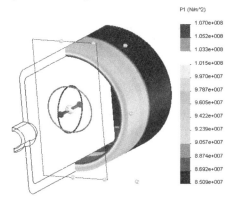

Figure 39 – Section view taken at the right plane.

4-37

15. Once again, to aid in examining stresses in the wall only, in the **Options** dialogue box, click to check ☑ **Plot on section only**. The resulting display shows stress contours on the cut section only. See Fig. 40.

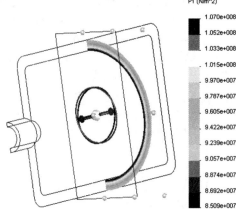

Figure 40 – Tangential stress display on mid-section of the model.

Figure 40 shows stresses on the cut section only and the section clipping drag-handles. The model is rotated to reveal the location of the section on an outline of the model.

16. In the **Section 1** dialogue box experiment by clicking the up ▲ or down ▼ arrows on the **Distance** spin box and observe the effect on stresses displayed in the model.

17. Click **[OK]** ✓ to close the **Section** property manager.

It is also possible to use the **Probe** feature, or other software options, in combination with any of the *Section* plots created above to further investigate stresses within a model. The next chapter extends current graphic display capabilities by introducing procedures to display and investigate results in a cylindrical coordinate system.

This concludes the analysis of thin and thick-wall pressure vessels using both *shell* and *solid* elements, respectively. Alternate ways of displaying graphical results were also explored.

It is suggested that the current model not be saved. Therefore, to exit COSMOSWorks and discard the solution for this model, proceed as follows.

1. Select **File** from the main menu at top of screen.

2. From the pull-down menu select **Close**. A **SolidWorks** window opens and prompts: **Save changes to Thick Wall Pressure Vessel?** Click **[NO]**.

> **Design Insight**
>
> When engaged in the design of any sort of pressure vessel, ASME Boiler and Pressure Vessel Codes[3] must be consulted.

---

[3] The ASME Boiler and Pressure Vessel Code, Sections I-XI, American Society of Mechanical Engineers, New York, NY, 1995.

EXERCISES

1. The cylinder pictured below allows for a maximum of three separate pipe connections. The pipe connection on the top cylindrical surface is considered **Immovable** due to its rigid external attachment (not shown). However, both pipe connections on the flat end are connected to flexible pipe segments (also not shown) that permit movement in the X (axial) and Y (vertical) directions. Movement in these directions is permitted to accommodate axial and circumferential expansion when the tank is pressurized. In other words, movement is only restricted in the Z-direction at these pipe connections. Perform a finite element analysis of this cylinder subject to the following guidelines. Open the file: **Pressure Vessel 4-1**.

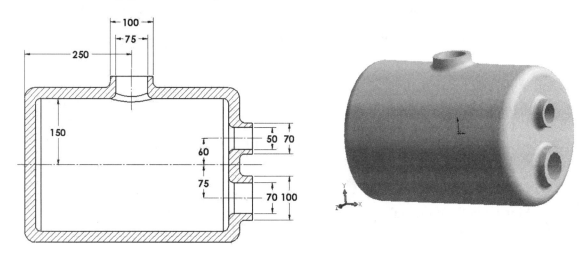

Figure E4-1 – Cylindrical pressure vessel with three pipe attachments. Wall thickness t = 25 mm.

- Material: **Alloy Steel** (Use S.I. units)

- Mesh: **High Quality** tetrahedral elements (NOTE: Determine whether $t \leq r_i/20$. Substituting yields, $25 \leq 150/20$. Since t is *not* $\leq r_i/20$, therefore thick-wall equations apply to classical calculations of stress in the cylinder walls.

- Internal pressure: **P = 14.5 MPa**, External pressure: Negligible

- Restraints: Pipe on top of cylinder is **Immovable**

Pipes on right end of the tank only restrain movement in the **Z**-direction. This restraint and its direction are somewhat arbitrary, but are included to demonstrate how alternative restraints might be applied. *Assistance dealing with these pipe restraints is included in the **Solution Guidance** box below.*

Apply **Symmetry** restraints on cut surfaces of the cylinder.

Analysis of Machine Elements using COSMOSWorks

Determine the following:
a. Develop a finite element model that includes: material selection, restraints, applied load, mesh the model (two elements are required across the cylinder wall thickness; one element thickness is permitted on pipe extensions), and a solution.

Solution Guidance

Begin by reducing the model to half of the tank geometry as follows.

- Toggle to SolidWorks and locate **Half-Model of Cylinder** in the SolidWorks manager tree; then **Unsuppress** it.
- Also while in the SolidWorks feature manager, defeature the model. **Supress** fillets at all pipe connections.

When assigning restraints on end surfaces of the horizontal pipes, proceed as follows.

i. Right-click **Load/Restraint** and from the pull-down menu, select **Restraints...** See Fig. E4-2.

ii. In top field of the **Type** dialogue box, select **On Flat Face** from the pull-down menu.

iii. Click the right *end face* of both pipe extensions located on the right-side of the cylinder. **Face<1>** and **Face<2>** appear in the **Planar Faces for Restraint** field.

iv. In the **Translations** dialogue box, set **Units** to **mm**.

Fig. E4-2 -Restraints applied to permit pipe movement in X and Y-directions, but prevent it in Z-direction.

v. Move the cursor over the second icon beneath the **Units** field. Its name should appear and identify the **Along Face Dir 2** field. When this icon is selected the field is highlighted (white) and the number "0" appears. This number indicates that zero translation is allowed in the selected direction.

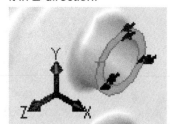

Fig. E4-3 – Restraint symbols in Z-direction.

Vectors displayed on the model should be directed in the Z-direction, see Fig. E4-3. Either the $\pm$ Z- direction is acceptable since displacement = 0.

vi. Click **[OK]** ✓ to close the **Restraint** property manager.

b. Create a stress contour plot of von Mises stress in the cylinder and all its connecting entities. Include automatic labeling of maximum von Mises stress on the plot. Does the maximum von Mises stress exceed material yield strength?

c. Use classical equations to calculate magnitudes of tangential and radial stresses at both the inside and outside surfaces of the cylinder.

d. Use finite element analysis to determine tangential stress in the cylinder wall opposite the top pipe connection. When determining this stress, reduce the range of stress values plotted to those slightly above and slightly below tangential stress magnitudes anticipated on the inside and outside surfaces. Plot the *appropriate stress* contours to show tangential stress in the cylinder wall.

e. Use the **Probe** feature to produce a graph of tangential stress variation through the cylinder wall from inside to outside.

f. Repeat step (d), but replace "tangential" stress with "radial" stress.

g. Repeat step (e), but replace "tangential" stress with "radial" stress.

h. Compare the percent difference between FEA results and results obtained from classical pressure vessel calculations determined in part (c). Results to be compared include the tangential stress at inside and outside surfaces of the cylinder and also radial stress at inside and outside surfaces. Use the following equation to compute the percent difference between results.

$$\% \text{ difference} = \frac{(\text{FEA result - classical result})}{\text{FEA result}} * 100 = \qquad [1]$$

2. The bracket shown below is one of two brackets used to attach a "carry all" to a motorcycle frame. **Chrome Stainless Steel** is used for the bracket material to enhance appearance and to provide corrosion resistance. Three bolts attach the bracket to the frame through holes on the vertical (left) leg at holes illustrated in Fig. E4-4. A downward design load of 18 lb is applied at the hole on tab **A**. This load represents the portion of load carried by tab **A** in a fully loaded "carry all." Perform a finite element analysis of this part using *shell* elements applied at the mid-surface of the part.

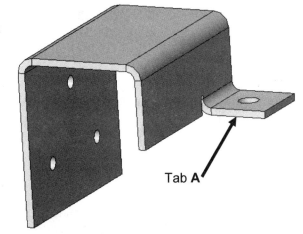

Figure E4-4 – Bracket used to attach a cargo case to a motorcycle frame.

Open the file: **Bracket 4-2**.

- Material: **Chrome Stainless Steel** (Use English units)

- Mesh: **High Quality** *shell* elements defined on the mid-surface. Use default mesh size. NOTE: For the **Shell mesh using mid-surface** option, thickness of the elements is automatically extracted from the SolidWorks model geometry.

- Restraints: Apply a **Fixed** restraint to the inner *surface* of the three bolt holes. Although the restraint is applied to a surface, it is transferred to the edge of shell element nodes.

- Force: **18 lb** applied vertically downward to the top surface of Tab A.

Determine the following:
a. Develop a finite element model that includes: material specification, restraints, applied force, mesh generation, and solution.

Solution Guidance

The procedure for applying a downward force on tab **A** is similar to that of applying a directed force to a split-line. Therefore, users are encouraged to apply the 18 lb downward force to the *edge* of the hole in tab **A** on their own. The following steps are provided in the event guidance is desired.

  i. Right-click **Load/Restraint** and from the pull-down menu, select **Force...** The **Force** property manager opens.

 ii. In the **Type** dialogue box, select ⊙ **Apply force/moment**.

iii. Also in the **Type** dialogue box the **Faces, Edges, Vertices, Reference points for Force** field is highlighted. Move the cursor onto tab **A** and select either the *edge* of the hole or the *inner surface* of the hole. **Edge<1>** or **Face<1>** is listed in the field depending upon which entity is chosen.

 iv. Click to highlight the **Face, Edge, Plane Axis for Direction** field, second field from top of the **Type** dialogue box, and select any vertical edge on the model. **Edge<1>** or **Edge<2>** appears in the field depending upon whether an edge or a surface was chosen in the preceding step.

  v. Verify that consistent **Units** are applied in the **Units** dialogue box.

 vi. In the **Force (per entity)** dialogue box, click to activate the **Along Edge** field. The field is highlighted (white) to indicate it is active. In this field type **18** to quantify the load applied to tab **A**.

> vii. Observe the direction of the 18 lb force. If not directed downward, check ☑ **Reverse direction**.
>
> viii. Click **[OK]** ✓ to close the **Force** property manager
>
> Return to complete the remainder of the solution on your own.

b. Plot von Mises stress on the top surface of the shell face. Consider the top surface to be that shown facing upward in Fig. E4-4 (i.e., the top side of the model is the side to which the external 18 lb force is applied). Include automatic labeling of maximum and minimum von Mises stress on this plot.

c. Repeat part (b) for the "bottom" surface of the bracket. By what percent do the maximum von Mises stresses on the top and bottom surfaces differ.

d. Question: On what surface does the maximum von Mises stress occur? Does this stress exceed the material yield strength? Describe factors, such as part geometry, location of the applied load, etc. that cause the maximum stress to occur where it does on the model.

> **Textbook Problems**
> In addition to the above exercises, it is highly recommended that additional problems involving pressure vessels and/or thin parts be worked from a design of machine elements textbook. Textbook problems provide a great way to discover errors made in formulating a finite element analysis because they typically are well defined problems for which the solution is known. Typical textbook problems, if well defined in advance, make an excellent source of solutions for comparison.

NOTES:

CHAPTER #5

INTERFERENCE FIT ANALYSIS

This example examines modeling of an interference fit between two mating parts. For purposes of discussion, the generic term "interference fit" is considered synonymous with the more specific terminology "force fit" or "shrink fit." However, COSMOSWorks documentation refers to all these fits as "Shrink Fits." These fit classifications are often used to join two members together without the need for other fastening devices such as set-screws or keys and keyways. Interference fit analysis is often considered together with pressure vessel analysis since the internal member exerts the equivalent of an outward pressure on the part into which it is force fit. And, conversely, the external member exerts the equivalent of an external pressure onto the part it surrounds.

Although this example solves an interference fit problem for a single set of part dimensions, it is important to recognize that, in practice, maximum and minimum stress levels occur due to tolerance variations of mating part dimensions. For example, a minimum level of interference must be ensured in order to transfer a desired torque between mating components. This case results when the smallest shaft is inserted into the largest hole. Conversely, when tolerances are such that the largest shaft is inserted into the smallest hole, maximum interference results in correspondingly higher stress levels. Another factor to consider is when different materials are used for mating parts. In these instances the effects of differential dimension changes due to thermal expansion or contraction may also have to be considered. The above scenarios are cited to alert the user to other factors that affect interference fits. However, because the emphasis of this text is on mastering COSMOSWorks software capabilities, these other factors are not considered in this example.

Learning Objectives
Upon completion of this example, users should be able to:

- Set-up and analyze an *interference fit* in an assembly.

- Define a *cylindrical* coordinate system.

- Examine results in a *local (cylindrical)* coordinate system.

- Generate a *report* summarizing results of a study.

- Apply alternative means of *controlling rigid body motion*.

Problem Statement
A partially dimensioned section view of a wheel and its axle are shown in Fig. 1. The wheel is part of an overhead traveling crane, which is used to transport heavy materials

from point-to-point on a factory floor. The 2.5041 inch diameter shaft is shrink fit into a wheel-hub with inside diameter of 2.5000 inches resulting in a 0.0041 inch diametral interference. This example is the first to involve analysis of an *assembly* of more than one part. Analysis of this model begins below.

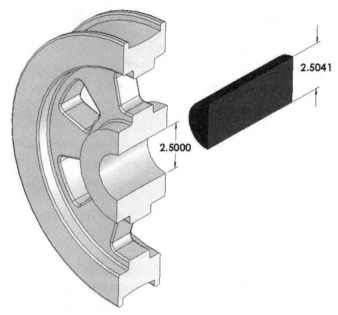

Figure 1 – Maximum shaft size and minimum hole size (minimum bore) for an interference fit between a shaft and crane wheel.

1. Open SolidWorks by making the following selections. (*Note:* "/" is used to separate successive menu selections.)

   Start / All Programs / SolidWorks 2008

2. When SolidWorks is open, select **File / Open** and open the COSMOSWorks file named **Wheel Shaft Assembly**. If the **Assembly** menu opens, click-and-drag it to a convenient location, perhaps on the menu bar, at the top of your screen.

Interference Check

Before beginning an analysis of this assembly, confirm that part interference does indeed exist. Proceed as follows to check for interference.

1. In the Main menu, click **Tools** and from the pull-down menu select **Interference Detection…**. The **Interference Detection** property manager opens as illustrated in Fig. 2.

2. In the **Selected Components** dialogue box observe that the **Wheel Shaft Assembly.SLDASM** is pre-selected as the assembly to be checked for interference. Also notice that the **Results** dialogue box currently indicates that interference is **Not calculated**.

3. In the **Selected Components** dialogue box, click the **[Calculate]** button. Immediately the *volume* of interference between the shaft and wheel is calculated and listed in the **Results** dialogue box as 0.06 in^3. Also, the interference region between the shaft and wheel is highlighted as illustrated in Fig. 2.

4. Click **[OK]** ✓ to close the **Interference Detection** property manager.

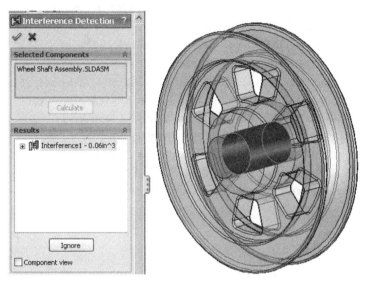

Figure 2 – The interference fit region is highlighted during an Interference Check of parts in the shaft and wheel assembly.

Creating a Static Analysis (Study)

1. Toggle to the **COSMOSWorks** analysis manager by clicking its icon located at the top-right side of the manager tree (circled in Fig. 3).

2. In the COSMOSWorks manager, Fig. 3, right-click **Wheel Shaft Assembly** and from the pull-down menu, select **Study…**. The **Study** property manager opens.

Figure 3 – Initial steps to create a Study in the COSMOSWorks Analysis Manager.

3. In the **Name** dialogue box type "**Force Fit Analysis**" to serve as a descriptive name for this Study.

4. Examine other fields within the **Study** property manager to ensure a **Solid Mesh** is selected and that **Static** is shown as the analysis **Type**.

5. Click **[OK]** ✓ to close the **Study** property manager.

Assign Material Properties to the Model

Unlike previous examples, materials for the **Shaft** and **Wheel** are *pre-assigned* within SolidWorks. This is observed by noting that a check mark "✓" appears on the **Solids** icon in the COSMOSWorks manager shown in Fig. 4. In this example, two different materials are defined. The shaft is made of **Alloy Steel** and the wheel material is **Cast Alloy Steel**. These material selections are automatically transferred into COSMOSWorks as part of the assembly definition created in SolidWorks. The next step describes how to view the material settings.

Figure 4 – Manager tree showing that materials for the Shaft and Wheel assembly are pre-selected in SolidWorks.

1. Click the "+" sign adjacent to the **Solids** icon. Then repeat this procedure for "+" signs adjacent to both the **Shaft-2** and the **wheel-1** icons. Next, move the cursor over the **Body 1** name for the shaft or the wheel in the manager tree to reveal the complete material designation. Figure 4 shows the complete name of the SolidWorks material specified for **Shaft-2**.

Before proceeding, verify that material properties are specified in English units as outlined below.

2. Beneath the **Shaft-2** icon, right-click **Body 1(Extrude1)** and from the pull-down menu select **Apply/Edit Material…**. The **Material** window opens. However, because material was specified in SolidWorks, most of the window is inactive (i.e., grayed out). Material properties appear in the right-half of the window.

3. Adjacent to **Units:** select **English (IPS)** units (if not already selected).

4. Click **[OK]** to close the **Material** window.

5. Repeat steps 2 through 4 for **Body 1(0.10 inch Fillets Rounds)** beneath the **wheel-1** icon. With the cursor resting on the above name, note that material for this part is listed as **(-[SW]Cast Alloy Steel-)**.

Analysis Insight
Material selection within SolidWorks is more limited than within COSMOSWorks. For this or other reasons, the procedure outlined in steps 1 to 4 above can be used to alter the material selection for either component by selecting alternate materials **From library files** as done in previous examples.

Interference Fit Analysis

Defeature the Model

An enlarged view of the shaft and wheel assembly is shown in Fig. 5. Careful observation reveals numerous small fillets and rounds on various edges of the model. Knowing in advance that the current assembly is to be analyzed using finite element methods, the developer of the original SolidWorks model expedited the defeaturing process by creating all the small fillets and rounds in a single step. Therefore, the defeaturing steps outlined below are very efficient.

Note: Larger fillets at the four "corners" of the spoke cut-outs are *not* removed from the model because, if removed, they might be a source of significant stress concentration. This judgment is based upon engineering insight. Proceed as follows to defeature the model.

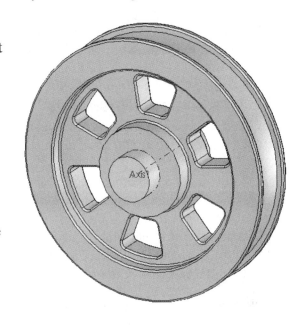

Figure 5 – Observe multiple fillets and rounds on various edges of the model.

1. Toggle back to SolidWorks by clicking the **SolidWorks** icon located at the top-left of the feature manager tree, shown boxed in Fig. 6.

2. Click the "+" sign adjacent to the **wheel<1>** folder, circled in Fig. 6, to display steps used to create the wheel model in SolidWorks.

3. Right-click the "**0.10 inch Fillets & Rounds**" highlighted near the bottom of the feature manger in Fig. 6. The pop-up menu, pictured to the right of the feature manager tree, also appears in Fig. 6.

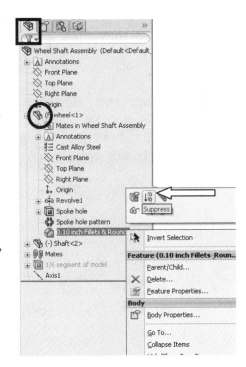

Figure 6 – Selections in the SolidWorks feature manager tree to simplify wheel geometry (i.e., defeature it).

4. Within the pop-up menu, select the **Suppress** icon to remove all small fillets and rounds from the wheel. This step simplifies wheel geometry prior to meshing the model.

Had other size fillets or rounds been specified on the model, the above process might have to be repeated multiple times. (Do *not* toggle back to COSMOSWorks at this time).

Apply Loads and Restraints

Unsuppress Part of the Model to Use Symmetry

Examination of the defeatured model, shown in Fig. 7, reveals symmetry in the repeated pattern of spokes and holes on the crane wheel. Once again, it is possible to reduce the solution size and speed up the solution process if model symmetry is used to advantage. Because there are six spokes and six openings, it is obvious that a 60° slice (360°/6 = 60°), or 1/6th of the model, is the smallest segment that should be used. It would be possible to select other fractions of the assembly for analysis as well. Other convenient segments would be 1/3, 1/2 or 2/3. Proceed as follows to apply the preceding symmetry observation.

Figure 7 – Observing symmetry of the defeatured shaft and wheel assembly.

1. Near the bottom of the SolidWorks manager tree, *left*-click the "**1/6 segment of model**" icon highlighted in Fig. 8. A pop-up menu, also shown in Fig. 8, appears on the feature manager.

2. From the pop-up menu, select the **Unsuppress** icon shown at the arrow on Fig. 8.

The result of steps 1 and 2 produces a 1/6th segment of the shaft and wheel assembly shown in Fig. 9.

3. Toggle back to the **COSMOSWorks** manager at this time.

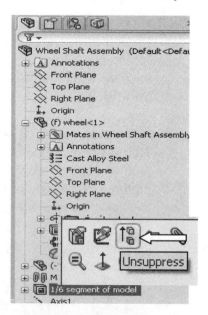

Figure 8 – Unsuppressing the 1/6th model of the shaft and wheel assembly.

Interference Fit Analysis

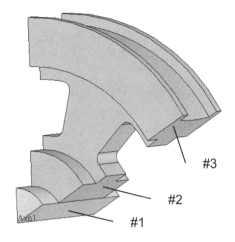

> **Analysis Insight**
> Due to model symmetry, it is possible to select a "slice" of the model that either intersects openings between wheel spokes (as illustrated in Fig. 9) or one that intersects spokes on both sides of an opening.
>
> The decision to model the segment shown in Fig. 9 is based on an understanding that a full picture of stresses in a spoke is more revealing than a split image of the stress distribution in half a spoke on either side of an opening.

Figure 9 – Symmetry applied to yield a 1/6th segment of the shaft and wheel assembly.

Define Symmetry Restraints

Symmetry restraints were first introduced in the previous chapter where segments of thin and thick-wall pressure vessels were considered. Symmetry restraints serve the purpose of making the cut portion of a symmetrical model behave *as if* the entire model were still present. Therefore, symmetry restraints must be applied to all surfaces created when the model was cut. Figure 9 shows one surface (#1) on the shaft and two surfaces (#2 and #3) on the hub and outer rim of the wheel, respectively. Of course, duplicate surfaces exist on the opposite side of the model. Therefore, symmetry restraints must be applied to a total of six surfaces. Proceed as follows to apply symmetry restraints.

1. In the COSMOSWorks analysis manager tree, right-click the **Load/Restraint** icon and from the pull-down menu select **Restraints…**. The **Restraint** property manager opens as shown in Fig. 10.

2. Beneath **Type**, open the pull-down menu and select the **Symmetry** restraint.

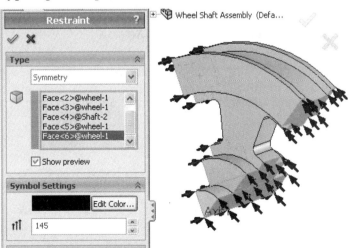

Figure 10 – **Symmetry** restraints applied to all cut surfaces of the shaft and wheel assembly.

3. The **Planar Faces for Restraint** field is highlighted (light blue) to indicate it is active and awaiting specification of surfaces to which symmetry restraints are to be applied. Move the cursor over the model and select the three numbered surfaces shown in Fig. 9. Also select their counterparts on the opposite side of the model. Rotate the assembly as necessary to select all faces.

With each selection, a new surface is highlighted and is listed as **Face<1>@wheel-1** (or **@Shaft-2** etc.), **Face<2>**, ... **Face<6>** in the **Planar Faces for Restraint** field. Also, restraint symbols appear on the model as illustrated in Fig. 10. (Symbol size was increased to enhance visibility).

4. Click **[OK]** ✓ to close the **Restraint** property manager. **Restraint-1** appears beneath the **Load/Restraint** icon in the COSMOSWorks manager tree.

Analysis Insight

As noted in the thick-wall pressure vessel example, while assignment of symmetry boundary conditions applies proper restraints to model missing portions of the part, those same boundary conditions may not provide sufficient restraints to prevent "rigid body motion" of the model.

For this example, note the following two facts: (a) All restraints shown in Fig. 10 are applied normal to the cut surfaces; and (b) *If* each restraint vector were broken into components, then each vector could be replaced by its X and Y components. Rotate and view the model to convince yourself of these two facts. Therefore, an obvious conclusion is that *no restraints* exist in the axial Z-direction. Thus, rigid body motion of the model is not restrained in the axial direction.

Because rigid body motion is not allowed in finite element studies, additional restraints must be added to the model as outlined next.

Apply Restraints to Eliminate Rigid Body Motion

Because the assembly consists of two parts, the shaft and the wheel, both parts must be restrained. The need to restrain both parts may seem counter-intuitive since a shrink fit is being defined between the shaft and wheel and a shrink fit is considered to be an immovable connection. However, this seeming contradiction provides an opportunity to note that shrink fit contact is considered frictionless by default within COSMOSWorks. Proceed as follows to apply restraints to prevent rigid body motion.

1. Right-click **Load/Restraint** and from the pull-down menu, select **Restraints...**. The **Restraint** property manager opens as shown on the left-side of Fig. 11.

2. Beneath **Type**, open the pull-down menu and select **Use Reference Geometry**. Reference geometry will be selected in a future step.

3. In the **Type** dialogue box the **Faces, Edges, Vertices for Restraint** field is active (light blue). Proceed to select two vertices on the model (one on the shaft the other on the wheel). Selected vertices are shown by two dots circled on Fig. 11. After selecting these vertices, **Vertex<1>@Shaft-2** and **Vertex<2>@wheel-1** appear in the active field. The actual number within < > depends on the order of selection.

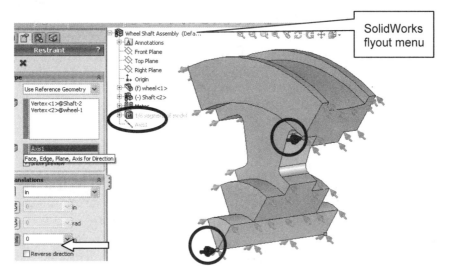

Figure 11 – **Restraint** property manager and selections of vertices and reference geometry needed to prevent rigid body motion of the shaft and wheel assembly.

4. Next, click to activate (light blue) the second field from top in the **Type** dialogue box. The identifier **Face, Edge, Plane, Axis for Direction** appears when the cursor is passed over this field as illustrated in Fig. 11.

5. Click to open the SolidWorks "flyout" menu, and select **Axis1** located at the bottom of this menu. **Axis1** is selected because it is oriented in the Z-direction, the direction in which rigid body translation is to be prevented. After making this selection, **Axis1** appears in the highlighted field. (Note: Your flyout menu may appear different than that shown in Fig. 11 depending upon what SolidWorks folders are open or closed). After selecting **Axis1**, close the SolidWorks flyout menu by selecting the "-" sign at top of the menu.

6. In the pull-down menu at top of the **Translations** dialogue box, set **Units** to **in**. Then, at the bottom of this dialogue box (see arrow in Fig. 11), select the **Axial** icon and accept the zero value shown. This value implies *no* translation is permitted in the **Axial** direction. Restraint symbols appear at the two locations in Fig. 11. Different colors and sizes are used to accentuate their presence.

7. Click **[OK]** ✓ to close the **Restraint** property manager. **Restraint-2** appears beneath the **Load/Restraint** icon in the COSMOSWorks manager.

8. Right-click **Load/Restraint** and from the pull-down menu select **Hide All**. This action temporarily hides restraint symbols to reduce clutter on the model.

The above steps ensure that the assembly is restrained against rigid body translations in the axial Z-direction.

> **Aside:**
> The focus of this example is on modeling and analyzing an interference fit. Therefore, no *external* loads or *reactions* are applied to the shaft and wheel assembly. External loads and reactions would be due to wheel-to-rail contact and shaft support reactions.
>
> As a consequence, only forces due to the interference fit are included. These forces are equal and opposite between contacting surfaces. In brief, forces caused by interference fits are internally balanced.

Define a Shrink Fit

Because the shaft outside diameter is larger than the wheel hole diameter, an interference fit exists between these two components. When analyzing shrink fits in COSMOSWorks, you are cautioned that the amount of interference should be greater than 0.1% of the larger diameter at the interface between mating parts. This restriction is imposed because the amount of overlap between mating parts should be sufficiently large to overcome approximations introduced during meshing. This is due to the fact that mesh size also includes a permissible tolerance range. If the amount of interference is too small, inaccurate solutions may result.

During the process of establishing a shrink fit between the shaft and wheel, it is necessary to specify the surfaces where interference occurs. To facilitate selecting these surfaces, components of the assembly are separated by creating an exploded view as follows.

1. From the Main menu select **Insert**. Then from the pull-down menu select **Exploded View…**. The **Explode** property manager opens as illustrated in Fig. 12.

Figure 12 – Manually creating an exploded view of the shaft and wheel assembly.

Interference Fit Analysis

Read the **How-To:** instructions at top of the **Explode** property manager and proceed as follows.

2. For this example, move the cursor onto the graphics screen and click to select the *shaft*. Immediately, an X, Y, Z coordinate triad appears to "float" near the shaft. Click-and-drag the vector aligned with the direction of the shaft axis and drag the shaft away from the wheel as illustrated in Fig. 12. This action is recorded in the **Explode Steps** table as **Explode Step1**.

3. Click **[OK]** ✓ to close the **Explode** property manager. Note: Either component could be moved in any direction to create an acceptable exploded view.

4. Within the COSMOSWorks manager tree, right-click the **Contact/Gaps (-Global: Bonded-)** icon. A pull-down menu appears. This is the first time this icon is used. It will be referred to simply as the **Contact/Gaps** icon.

5. From the pull-down menu select **Define Contact Set…**. The **Contact Set** property manager opens as illustrated in Fig. 13.

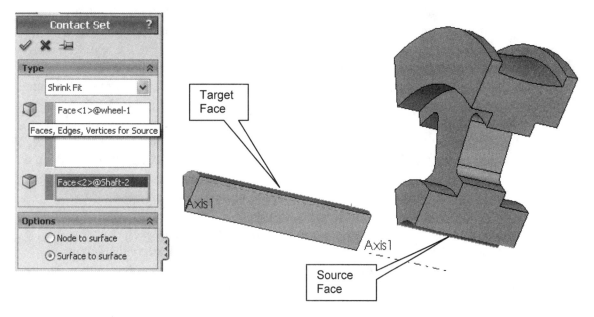

Figure 13 – Defining faces where a shrink fit occurs between the shaft and wheel. For this example the **Target** and **Source** faces can be interchanged without consequence.

6. At the top of the **Type** dialogue box, click to open the pull-down menu and select **Shrink Fit** as the type of contact to be modeled between the shaft and wheel.

7. The **Faces, Edges, Vertices for Source** field, located beneath the pull-down menu, is activated (light blue) and awaits input. Rotate the model as necessary and select the *inside* surface of the hole as the "source." **Face<1>@wheel-1** appears in the active field of Fig. 13.

5-11

8. Next, click to activate the second field beneath the pull-down menu. This is the **Faces for Target** field. Move the cursor over the model and select the *cylindrical* surface of the shaft as the "target," also shown in Fig. 13. **Face<2>@Shaft-2** appears in the active "target" field. NOTE: In this example either surface can be designated as the "source" or the "target." However, this observation is not valid in all cases. See item (c) in the Analysis Insight section below for more details about "source" and "target" designation.

Analysis Insight:

a) Before closing the **Contact Set** dialogue box, note the **Friction:** dialogue box at the bottom of this property manager. As noted earlier, interference fits are considered frictionless by default. However, if our goal were to investigate torque transmitted by the interference fit joint, then friction between mating surfaces could be defined here. This example does not examine **Friction:** effects.

b) Also before closing the **Contact Set** dialogue box, briefly revisit the pull-down menu that lists the various types of **Contact/Gap** elements available within COSMOSWorks. Understanding these criteria and a built-in hierarchy that accompanies them is essential to proper modeling for a variety of design and analysis situations. For example, in *assemblies* of multiple parts, COSMOSWorks applies a system default assumption that all contacting surfaces are **Bonded** together. This assumption treats all contacting faces of different components in an assembly as if they are permanently "glued" together. As such, the **Bonded** condition is applied *Globally* to all contacting parts. This assumption is applied unless one of the other contact conditions, such as the shrink fit used in this example, is specified *locally* (i.e., between two or more contacting components in a larger assembly). Contact conditions defined in the **Contact Set** property manager have higher priority than do Global or component contact definitions.

To pursue this topic further, the interested reader is referred to **Help** located on the COSMOSWorks main menu. From the **Help** pull-down menu, select **COSMOSWorks Help Topics** and type "**contact gap**." Review the multiple help topics listed there. Virtually all types of **Contact/Gap** elements found there are excellently described and illustrated by animated examples. Kurowski's[1] text summarizes these contact sets in tabular form.

c) The use of "source" and "target" is interchangeable in this example since both contacting surfaces are *faces*. However, the true definition of a "source" can refer to faces, edges, or vertices. Whereas the true definition of a "target" *only* refers to a face. Thus, in terms of the shrink fit modeled here, contacting faces satisfy the criteria for both a "source" and a "target." On the other hand, if a specific contact were to be defined between the surface of the shaft and the *edge* of the wheel hub, then the shaft *face* must be the "target" while the hub *edge* must be the "source."

---

[1] Kurowski, Paul M., Engineering Analysis with COSMOSWorks Professional, SDC Publications, 2008.

Interference Fit Analysis

9. Finally, in the **Options** dialogue box select ⊙ **Surface to surface**, then click **[OK]** ✓ to close the **Contact Set** property manager. After closing this property manager, **Contact Set-1 (-Shrink fit <wheel-1, Shaft-2>-)** appears in the COSMOSWorks manager tree below the **Contact/Gaps** icon.

Mesh the Model and Run the Solution

Although a correct solution is obtained with the model in either the exploded or unexploded state, return it to its *un*exploded state before meshing the model. This is done to permit observation of node alignment, or lack thereof, between nodes on the shaft and wheel after meshing. Un-exploding the model is accomplished as follows.

1. Click the **Configuration Manager** icon, located adjacent to the COSMOSWorks icon, shown circled in Fig. 14.

2. Within the Configuration manager, click the "+" signs adjacent to + **Wheel Shaft Assembly Configuration(s)** and adjacent to + **Default <Display State-1> [Wheel Shaft Assembly]**, (if not already selected).

3. Right-click **ExplView1**, then from the pull-down menu shown in Fig. 14, select **Collapse** to return the model to its assembled view.

Figure 14 – Accessing the **Configuration Manager** tree to "collapse" (*un*exploded) the shaft and wheel assembly.

4. Toggle back to the **COSMOSWorks Analysis Manager** icon.

5. In the COSMOSWorks manager tree, right-click the **Mesh** icon and from the pop-up menu select **Create Mesh…**. The **Mesh** property manager opens.

6. Click to open the **Options** dialogue box. Verify that mesh **Quality** is set to ⊙ **High**. Make no other changes on this tab.

7. Briefly return to the **Mesh Parameters** dialogue box at top of the **Mesh** property manager and verify that inches (**in**) are shown in the **Units** field.

8. The default mesh size is considered acceptable for this analysis, therefore, proceed *directly* to the Solution. This action has not yet been illustrated in this text. Do this by clicking the check-box adjacent to ☑ **Run analysis after meshing** located near the bottom of the **Mesh Parameters** dialogue box.

9. Finally, click **[OK]** ✓ to close the **Mesh** property manager. Meshing begins automatically and is followed immediately by the Solution. The solution process takes slightly longer than when an interference fit is not specified (approximately 30 seconds on a Pentium 4 PC).

10. If the mesh is not displayed after the solution, right-click the **Mesh** icon and from the pull-down menu, select **ShowMesh**.

The meshed model appears in Fig. 15. Zoom in to examine nodes on the contacting surfaces circled in this figure. Notice that nodes on the shaft and wheel are not necessarily aligned with one another.

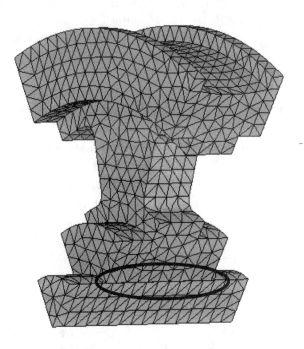

Figure 15 – Meshed model of the shaft and wheel assembly. Approximately eleven nodes exist along the contacting surfaces circled above.

Examination of Results

Default Stress Plot

Begin by examining the vonMises stress plot to gain an overview of results. Desired appearance attributes for this plot are defined below so that they can be copied when producing additional plots.

1. If necessary, click the "+" sign adjacent to the **Results** folder to display a list of the three default plots, **Stress1**, **Displacement1** and **Strain1**.

2. Double-click **Stress1 (-vonMises-)** to display this plot. Zoom-in on the model to observe deformation between the shaft and wheel hub in the interference fit region. If the deformed model does *not* appear or if units on the stress legend are *not* psi, proceed to the next step. Otherwise skip to step 6.

3. Right-click **Stress1 (-vonMises-)** and from the pop-up menu, select **Edit Definition…**. The **Stress Plot** property manager opens.

4. In the **Display** dialogue box, change **Units** to **psi**.

5. In the **Deformed Shape** dialogue box, check ☑ **Deformed Shape** (if not already selected) and click to select ⊙ **Automatic**. Then click **[OK]** ✓ to close the **Stress Plot** property manager. The system default, exaggerated deformation scale, is applied to an exploded view of the image.

6. After briefly examining the deformed model, right-click **Stress1 (-vonMises-)** and from the pop-up menu, select **Edit Definition…**. The **Stress Plot** property manager opens.

7. In the **Deformed Shape** dialogue box, select ⊙ **True Scale**. This action displays distortions between the shaft and wheel parts at true scale (1:1).

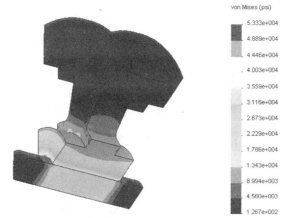

Figure 16 – vonMises stress plot incorporating modified display characteristics outlined in this section.

Next, adjust appearance of the von Mises plot so that, at the conclusion of this process, the plot appears as shown in Fig. 16. Proceed as follows.

8. Click ⌄ to open the **Advanced Options** dialogue box. Then, click to *clear* the check from ☐ **Average results across boundary for parts**. This action is *very important* because parts of this assembly are made of two *different materials*. Thus, it is not appropriate to average stress results across these two different parts.

9. Also within the **Stress Plot** property manager, click ⌄ to open the **Property** dialogue box. In this dialogue box, click to "check" ☑ **Include title text** and in the empty field, type your name followed by a descriptive title for the plot.

10. Click **[OK]** ✓ to close the **Stress Plot** property manager.

11. Once again, right-click **Stress1 (-vonMises-)** and from the pop-up menu select **Settings…**. The **Settings** property manager opens.

12. In the **Fringe Options** dialogue box, click the pull-down menu and select **Discrete** as the display mode for fringes.

13. Within the **Boundary Options** dialogue box, open the pull-down menu and select **Model** to display a black outline on the model. (This item is a user preference).

14. Click **[OK]** ✓ to close the **Settings** property manager.

Observation of Fig. 16 clearly reveals that stress levels in the vicinity of the shrink fit (53,330 psi) *exceed* Yield Strength of the wheel material, which is 35000 psi. Figure 17 shows a summary of properties for the wheel in the **Material Details** window. The material yield strength **SIGYLD** = 34994 psi ≈ 35000 psi is listed adjacent to the arrow. Another way to examine this fact is explored in the "Aside" section below.

Aside:
Recall that material information is available at any time by clicking the "+" sign adjacent to the **Solids** folder to reveal **Shaft-2** and **wheel-1**. Next, click the "+" adjacent to **wheel-1** to reveal **Body 1(0.10 inch Fillets & Rounds) [-(SW)Cast Alloy Steel-)**. Right-click this name and in the pull-down menu, shown in Fig. 17, and select **Details...** This action opens the **Material Details** window located at the right side of the Fig. 17.

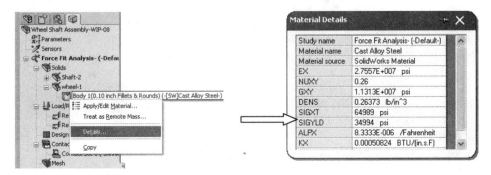

Figure 17 – Accessing material property details for the wheel to permit comparison of material Yield Strength with resulting stress in the component.

Notice that the material yield strength does not appear on the vonMises stress plot as occurred in earlier examples. The reason for this is that assemblies involve multiple parts, which are often made of different materials. Therefore, the potential for different material properties prohibits the display of any single yield strength value. Because of this, an alternative procedure to that of using the *Safety Factor* check can be applied to display regions where yield strength is exceeded. To apply this method, begin by making a *temporary* change in the **Chart Options** property manager outlined below.

a) Right-click **Stress1 (-vonMises-)** and from the pull-down menu select **Show**.

b) Right-click **Stress1 (-vonMises-)** and from the pull-down menu select **Chart Options...** The **Chart Options** property manager opens.

c) Near the bottom of the **Display Options** dialogue box click to select ⦿**Defined:**. Then, in the bottom box (corresponding to maximum stress) type a value somewhat greater than the yield strength for the wheel as **38000** (or **3.8e4**).

d) Do *not* close the **Chart Options** property manager at this time.

A new plot of vonMises stress within the shaft and wheel assembly is shown in Fig. 18. However, this plot highlights (in red) the entire region where stress due to the interference fit exceeds the yield strength. Examination of the colored stress chart reveals that stresses above approximately 34840 psi are shown in red. Therefore, the stress region above the yield strength (35,000 psi) is easily identified. Next, reset to the original plot settings as follows.

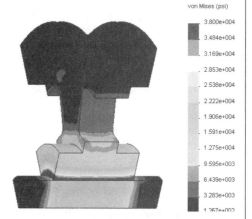

Figure 18 – Locating regions where yield strength is exceeded by re-defining chart **Display Options**.

e) Return to the middle of the **Display Options** dialogue box and select ⊙ **Automatic**.

f) Click **[OK]** ✓ to close the **Chart Options** property manager. The image is reset to display the full-range of von Mises stress throughout the assembly.

The fact that yield strength of the wheel is exceeded may be cause for redesign of the interference fit and/or selection of an alternative wheel material. However, since the goal of this example is to focus on defining interference fits and interpreting results of those fits, a re-design of the shaft and wheel assembly is not pursued here. We now return to investigate alternative means to display results.

Stress Plots in the Cylindrical Coordinate System

Circumferential (Tangential or Hoop) Stress

Circumferential (or) hoop stress is one of two stresses easily calculated using classical stress equations for a shrink fit. Because traditional stresses like σ_X, σ_Y, σ_Z, σ_1, σ_2, σ_3, τ_{xy}, τ_{yz}, τ_{zx}, etc., in a Cartesian (X, Y, Z) coordinate system are not conducive for representing circumferential stress in cylindrical parts, a cylindrical coordinate system is introduced below to facilitate the examination of results.

When establishing a cylindrical coordinate system, *any* axis can be used to define the axis of a local cylindrical coordinate system. However, once an axis is selected, the stresses usually associated with SX = σ_X, SY = σ_Y, and SZ = σ_Z take on new meanings as summarized in Table 1.

Analysis of Machine Elements using COSMOSWorks

Table 1 – Correspondence between stresses in a Cartesian coordinate system and their equivalents in a cylindrical coordinate system.

Original Meaning of Stress	New Meaning in the Cylindrical Coordinate System
SX = stress in X-direction σ_x.	SX = stress in *radial* direction relative to the selected reference axis (σ_r).
SY = stress in Y-direction σ_y.	SY = stress in *circumferential* directio From the pull-down menu select **Close**. n relative to the selected reference axis (σ_t).
SZ = stress in Z-direction σ_z.	SZ = stress in *axial* direction relative to the selected reference axis (σ_a).

In brief, no matter what axis is chosen to be the reference axis[2] of a cylindrical coordinate system, the correspondence between radial, circumferential, and axial stresses listed in Table 1 remains valid.

Before selecting an axis to define a cylindrical coordinate system, a *copy* of the existing vonMises stress plot is made. The primary reason for making a copy of the stress plot is to save the time it would take to recreate all of the graphic settings defined above. Copy the plot as follows.

1. Right-click **Stress1 (-vonMises-)** and from the pop-up menu select **Copy**.

2. Right-click the **Results** folder and from the pop-up menu select **Paste**. A plot named **Copy[1] Stress1 (-vonMises-)** is added to the list of plots beneath the **Results** folder. Recall that an alternative means of making this copy is to click-and-drag **Stress1 (-vonMises-)** onto the **Results** folder.

3. Double-click **Copy[1] Stress1 (-vonMises-)** to display an *identical* plot of the vonMises stress including all previously defined plot display options.

Next, a cylindrical coordinate system is defined. **Axis1**, aligned with the shaft centerline, is the logical choice for the axis of a cylindrical coordinate system. However, since the goal is to display circumferential (hoop) stress in the cylindrical coordinate system, two changes must be made to the *copied* plot. First, **SY** must be specified as the stress to be viewed in this plot because, according to Table 1, **SY** = *circumferential* stress in a cylindrical coordinate system. And second, **Axis1** must be defined as the axis of a cylindrical coordinate system. Proceed as follows to alter the copied plot.

4. Right-click **Copy[1] Stress1 (-vonMises-)** and from the pop-up menu select **Edit Definition...** The **Stress Plot** property manager opens as shown in Fig. 19.

5. In the **Display** dialogue box, click to open the stress **Component** pull-down menu and change it to **SY: Y Normal Stress**. This action also opens the **Advanced Options** dialogue box also shown in Fig. 19. Recall that according to Table 1, stress **SY** represents *circumferential* stress in a cylindrical coordinate system

---

[2] Reference axis refers to the axis aligned with the centerline of the cylindrical coordinate system.

6. Move the cursor over the top field in the **Advanced Options** dialogue box to identify it as the **Plane, Axis, or Coordinate System**. This field is highlighted (light blue) to indicate it is active and awaiting user input.

7. In the graphics screen, either click to select the center-line of the shaft axis (or) click to open the SolidWorks flyout menu and select **Axis1** shown circled on Fig. 19. This selection establishes **Axis1** as the reference axis of the cylindrical coordinate system. The name **Axis1** appears in the top field of the **Advanced Options** dialogue box.

8. Verify that ⊙ **Node Values** is selected and click **[OK]** ✓ to close the **Stress Plot** property manager.

Figure 19 – Selections to define a cylindrical coordinate system and circumferential stress.

The circumferential stress distribution, as it appears in the cylindrical coordinate system, is shown in Fig. 20.

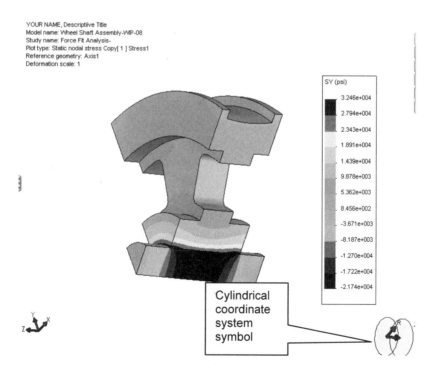

Figure 20 – Illustration showing circumferential stress distribution in the shaft and wheel in a cylindrical coordinate system.

5-19

9. Once again move the cursor onto **Copy[1] Stress1 (-Y normal-)** and click-*pause*-click to select this name. Then within the name field, type "**Circumferential Stress**" and press **[Enter]**. The plot name is changed to **Circumferential Stress (-Y normal-)**.

Observations:

Figure 20 displays the circumferential stress distribution in a cylindrical coordinate system. Circumferential stress in the wheel hub is analogous to circumferential (tangential or hoop) stress induced in a thick-wall cylinder subject to internal pressure. In this example, however, an interference fit is the cause of internal pressure on the external wheel hub. In Fig. 20, and on your screen, notice that the Global X, Y, Z coordinate system triad is supplemented by a cylindrical coordinate system symbol.

Also, observation of stress magnitude reveals that maximum circumferential stress does *not* exceed the wheel material yield strength. However, recall that vonMises stress of Fig. 16 and given by equation [1], is made up of *all* components of principal stress at a point. For simplicity, consider only a two-dimensional state of stress within the wheel hub, then both circumferential *and* radial stress components contribute to vonMises stress (σ') as follows.

$$\sigma' = \sqrt{\sigma_t^2 - \sigma_t \sigma_r + \sigma_r^2} \qquad [1]$$

where - σ_t = circumferential stress (tangential or hoop stress)
σ_r = radial stress

This brief digression emphasizes the need to consider the *appropriate* stress when reaching conclusions about part safety.

Radial Stress

The above observation leads to the conclusion that *radial* stress must also be examined in order to obtain a complete picture of stress due to the interference fit between the shaft and wheel assembly. Proceed as follows to produce a plot of radial stress.

1. Make a copy of **Circumferential Stress (-Y normal-)** by following either procedure outlined in steps 1 and 2 of the previous section. Try this on your own. The new copy is listed as **Copy[1] Circumferential Stress (-Y normal-)** beneath the **Results** folder.

2. Double-click **Copy[1] Circumferential Stress (-Y normal-)** to open a new copy of the previous (identical) plot.

3. Right-click **Copy[1] Circumferential Stress (-Y normal-)** and from the pull-down menu select **Edit Definition…**. The **Stress Plot** property manager opens as shown in Fig. 21.

Because the preceding steps copied a plot that already includes a cylindrical coordinate system, there is no need to repeat steps required to define that coordinate system. Verify this by observing that **Axis1** appears in the top field of the **Advanced Options** dialogue box shown in Fig. 21.

4. In the **Display** dialogue box, click to open the **Component** pull-down menu and from the stresses listed, select **SX: X Normal Stress** which, according to Table 1, represents the *radial* stress in a cylindrical coordinate system.

5. Click **[OK]** ✓ to close the **Stress Plot** property manager.

Before proceeding, change the name of the plot currently named **Copy[1] Circumferential Stress (-X normal-)** to reflect that it now contains a plot of *radial* stress. Try this on your own or see the following step.

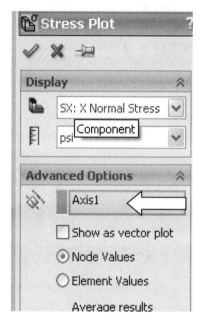

Figure 21 – Selecting SX: X Normal to display *radial* stress in the cylindrical coordinate system.

6. Click-*pause*-click the current folder and type **Radial Stress**. Then press **[Enter]**. The revised plot name should appear as **Radial Stress (-X normal-)**.

The resulting plot of radial stress is shown in Fig. 22. Radial stress represents stress normal to the two contacting surfaces. As such, radial stress represents the contact pressure between the shaft and wheel bore. Also observe the ± sign of radial stress magnitudes. Virtually all radial stress magnitudes are negative. Negative signs indicate compressive stresses acting on the outer surface of the shaft and on the inner surface of the wheel hub in the vicinity of the interference fit. This result is consistent with a general understanding of what happens in a shrink fit.

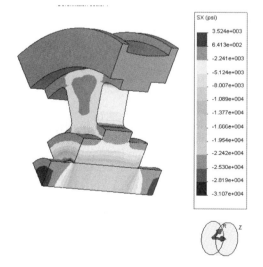

Figure 22 – Plot of *radial* stress distribution throughout the shaft and wheel assembly caused by the interference fit.

Verification of Results

Stress Predicted by Classical Interference Fit Equations

Consistent with previous examples, a check is made to determine validity of results of the current analysis. However, classical equations for interference fits between mating parts assume that (a) both components are of equal length, and (b) components are uniform thickness throughout their contact length. Both of these assumptions are violated in the current example, therefore, some differences between classical and FEA results are anticipated. The next paragraph describes two reasons for the expected differences.

First, the fact that both inner and outer members are not of equal length gives rise to higher stresses near both ends of the wheel hub at locations indicated in Fig. 23. These regions of high stress are caused by stress concentration effects due to pressure induced by the shrink fit combined with the geometric discontinuity where the shaft meets the hub. Higher tensile stress in this region of the hub was previously observed in Figs. 16 and 20. Second, a stiffening effect occurs in the central region of the hub, circled in Fig. 23, where the hub is thicker due to spokes and a built-up section located near its center. Stiffening in this region has the effect of altering magnitudes of all stresses examined thus far (hoop stress, vonMises stress, and radial stress).

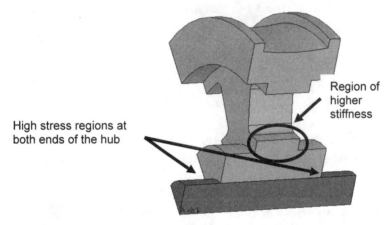

Figure 23 – Geometric differences between the actual model and assumptions of classical interference fit equations.

Classical interference fit equations are applied below to determine circumferential and radial stresses at the contact surface. Nominal part dimensions, given in Fig. 1, combined with SolidWorks material properties for the shaft and wheel are used in equation [2] to determine the common contact pressure "p" between mating parts.

Solution for common contact pressure between force fit parts.

$$\delta = \frac{pR}{E_o}\left(\frac{r_o^2 + R^2}{r_o^2 - R^2} + v_o\right) + \frac{pR}{E_i}\left(\frac{R^2 + r_i^2}{R^2 - r_i^2} - v_i\right)$$

[2]

$$0.00205 = \frac{p(1.25)}{27.6e6}\left(\frac{(2.5)^2 + (1.25)^2}{(2.5)^2 - (1.25)^2} + 0.26\right) + \frac{p(1.25)}{30.5e6}\left(\frac{(1.25)^2 + 0^2}{(1.25)^2 - 0^2} - 0.28\right)$$

Solving equation [2] for contact pressure between the wheel and shaft yields
p = 17,530 psi = radial stress at the contact surface = σ_r.

> Where: δ = radial interference
> p = the *unknown* contact pressure between mating parts
> R = common radius at contacting surfaces
> r_i = inside radius of inner member (r_i = 0 for a solid shaft)
> r_o = outside radius of external member (wheel hub radius is used)
> E_i = modulus of elasticity for inner member (shaft)
> E_o = modulus of elasticity for outer member (wheel)
> v_i = Poisson's ratio for inner member (shaft)
> v_o = Poisson's ratio for external member (wheel)

Next, using the contact pressure "p" that exists between the shaft and wheel hub, circumferential stress at the outer surface of the inner member (i.e., on the shaft surface) is computed in equation [3].

$$(\sigma_t)_i = -p\frac{R^2 + r_i^2}{R^2 - r_i^2} = -17530\left(\frac{(1.25)^2 + 0^2}{(1.25)^2 - 0^2}\right) = -17,530\,psi$$

[3]

Finally, the circumferential stress at the inner surface of the outer member (i.e., circumferential stress on inner surface of the wheel hub) is given by equation [4].

$$(\sigma_t)_o = p\frac{r_o^2 + R^2}{r_o^2 - R^2} = 17530\left(\frac{(2.5)^2 + (1.25)^2}{(2.5)^2 - (1.25)^2}\right) = 29,200\,psi$$

[4]

Stress Predicted by Finite Element Analysis

Radial Stress Comparison

Given the prior observations about obvious differences between classical equation assumptions and the actual model, it is logical to expect differences between results predicted by classical equations and the finite element analysis (FEA). Although finite element stress plots were already obtained for the entire assembly, this section focuses on results at contact locations predicted by classical equations [2] through [4]. Proceed as follows to determine radial stress (i.e., the contact pressure) between the shaft and wheel.

1. From the main menu select **Insert** and from the pull-down menu select **Exploded View…**. The **Explode** property manager opens.

Since an exploded view was previously defined for this assembly, it is displayed upon making the above selection.

2. Click **[OK]** ✓ to close the **Explode** property manager.

3. In the COSMOSWorks manager, right-click **Radial Stress (-X normal-)** and from the pull-down menu choose **List Selected**. The **Probe Result** window opens as shown in Fig. 24.

4. In the **Options** dialogue box, select ⊙ **On selected entities**.

5. In the **Results** dialogue box the **Faces, Edges, Vertices** field is highlighted and awaiting selection of the item for which results are to be displayed. On the graphics screen, rotate and zoom-in on the inner surface of the wheel hub shown "boxed" in Fig. 24, then click to select it. **Face<1>@wheel-1** appears in the field.

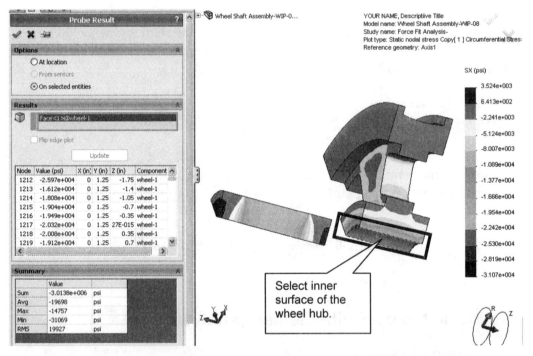

Figure 24 – Inside of the wheel surface is selected to examine results of all radial stresses acting on that surface. Results are summarized in the **Probe Result** table.

6. In the **Results** dialogue box, click the **[Update]** button.

Immediately the table is populated with data. The **Value (psi)** column contains values of *radial* stress at all nodes on the inside surface of the wheel hub. Also included are node numbers, X, Y, Z locations of each data point, and a column identifying the **Components**, in this case **wheel-1**, to which the data applies. NOTE: To accurately view tabulated data it may be necessary to click-and-drag the right margin of the manager tree as well as column-edges within the table to increase their width.

In the **Summary** dialogue box, located at the bottom of the **Probe Result** property manager, observe that radial stress values on the selected surface are summarized in several different forms (**Sum**, **Avg**, **Max**, **Min** and **RMS**). For this example, consider only the **Avg** value listed as **-19698 psi** for the radial stress. Using this value, a comparison with the result of classical equation [2] above yields:

$$\% \text{ difference} = \frac{\text{FEA result - classical result}}{\text{FEA result}} * 100 = \frac{19698 - 17530}{19698} * 100 = 11.0\% \quad [5]$$

While the percent error found in equation [5] exceeds *desired* expectations, it is probably reasonable given the differences between assumptions associated with use of classical equations and the actual model geometry. This larger than desired difference is clearly a case where St. Venant's principle, once again, influences the final results.

7. Select the **[OK]** ✓ to close the **Probe Result** property manager.

Circumferential Stress Comparison

Next return to the **Circumferential Stress (-Y normal-)** plot and repeat the above procedure, except this time determine the circumferential stress on the same inner surface of the wheel hub. Try this on your own. If needed, the following steps provide an outline of necessary steps.

1. Right-click **Circumferential Stress (-Y normal-)**, then from the pull-down menu select **Show**. Alternately, double-click **Circumferential Stress (-Y normal-)**.

2. Right-click **Circumferential Stress (-Y normal-)** and from the pull-down menu choose **List Selected**. The **Probe Result** property manager opens.

3. In the **Options** dialogue box, select ⊙ **On selected entities**.

4. On the graphics screen, click to select the inner surface of the wheel hub shown "boxed" in Fig. 24.

5. In the **Results** dialogue box, click the **[Update]** button.

Compare the **Avg** value for the circumferential stress from the finite element analysis with that predicted by equation [4] above. The comparison is shown in equation [6] below.

$$\% \text{ difference} = \frac{\text{FEA result - classical result}}{\text{FEA result}} * 100 = \frac{28845 - 29200}{28845} * 100 = 1.23 \% \quad [6]$$

6. Click **[OK]** ✓ to close the **Probe Result** property manager.

These results represent much better agreement between the Finite Element Analysis stress prediction and the classical equation solution. It might appear that the next logical step would be to compare finite element results for circumferential stress on the outer surface of the shaft with results predicted by equation [3]. However, because the process outlined above selects data at *all* nodes on the selected surface, poor agreement would result because a considerable area, on both ends of the shaft surface, lies outside the contact area defined by the interference fit. For this reason, these results are not compared here. However, if desired, *Split Lines* could be added to the shaft to limit comparison to the contact region only.

Quantifying Radial Displacements

Through additional application of the **List Selected** feature, COSMOSWorks provides a convenient means to verify the interference fit imposed on the mating parts of this example. Note: Because the shaft and wheel hub are of different lengths, results obtained from the following analysis will differ from those predicted by classical equations. However, in instances where inner and outer members are of the same length, such as Exercise 1 at the end of this chapter, excellent agreement of results is obtained. Proceed as follows to determine the deformation of each part.

1. In the COSMOSWorks manager, right-click the **Displacement1 (-Res disp-)** plot icon, and from the pull-down menu, select **Show**. The **Displacement** plot is displayed.

2. Again, right-click **Displacement1 (-Res disp-)** and from the pull-down menu, select **Edit Definition…** The **Displacement Plot** property manager opens as shown in Fig. 25.

3. In the **Component** field of the **Display** property manager, select **UX: X Displacement**. According to Table 1, **UX: X Displacement** corresponds to *radial* displacement in a cylindrical coordinate system.

4. Click to open the **Advanced Options** dialogue box. The **Plane, Axis or Coordinate System** field is highlighted (light blue) and awaits selection of an axis to define a cylindrical coordinate system.

5. Open the SolidWorks flyout menu (if not already open) and click to select **Axis1** to define it as the reference axis for a cylindrical coordinate system. **Axis1** should appear in the highlighted field.

6. In the **Display** dialogue box, verify that **Units** are set to **in** and that ⦿ **True scale** is selected in the **Deformed Shape** dialogue box.

7. Click **[OK]** ✓ to close the **Displacement Plot** property manager.

8. If the plot produced in the preceding step does not appear on the graphics screen, double-click **Displacement1 (-X disp-)** to display the plot.

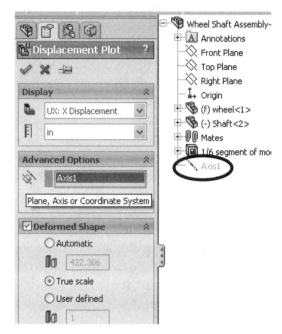

Figure 25 – Set-up of the **Displacement Plot** property manager to produce a plot of radial displacement.

9. Right-click **Displacement1 (-X disp-)** and from the pull-down menu choose **List Selected**. The **Probe Result** property manager opens.

10. In the graphics area, click the inner surface of the wheel hub shown "boxed" in Fig. 24. Then in the **Results** dialogue box, click the **[Update]** button.

The preceding step causes radial displacements at every node on the selected surface to be listed in the table of the **Results** dialogue box. Once again it may be necessary to click-and-drag to adjust column widths so that complete numerical values of displacement can be observed.

Of primary importance to this analysis is the **Avg** value of displacement listed in the **Summary** dialogue box. The value shown there corresponds to *radial* displacement of the inner surface of the wheel hub after the interference fit is accomplished. The **Avg** value of deformation on the inside surface of the wheel is listed as **0.0015202 in**. Without exiting the **Probe Result** property manager, proceed to examine deformation on the shaft surface as follows.

11. In the **Faces, Edges or Vertices** field, top field in the **Results** dialogue box, right-click **Face<1>@wheel-1** and from the pop-up menu select **Delete**.

12. Next, rotate the model so that the cylindrical surface on top of the shaft is visible and click to select it. **Face<2>@Shaft-2** appears in the active field.

13. Once again, click the **[Update]** button in the **Results** dialogue box. The table is now populated with values of the *radial* displacement on the outer surface of **Shaft-2**.

14. Notice the **Avg** value of radial displacement listed in the **Summary** table is **-0.00031518 in**.

15. Exit from the **Probe Result** property manager by clicking **[OK]**✓.

To determine total radial interference between the shaft and wheel, the absolute values of the above two **Avg** radial displacements are added together in equation [7]. The sum of these two values is approximately equal to the original radial interference specified in the opening statement of this example.

$$\text{Total radial interference} = |0.0015202| + |-0.00031518| = 0.0018354 \text{ in} \quad [7]$$

And, from the given information, page 5-2, for this example,

$$\text{radial interference} = \tfrac{1}{2} * \text{diametral interference} = \tfrac{1}{2} * 0.0041 = 0.00205 \text{ in} \quad [8]$$

The difference between total radial interference for equation [7], based on displacement plots, and the given interference from equation [8] is,

$$\text{Difference} = 0.00205 - 0.00184 = 0.00021 \text{ in (rounded values)} \quad [9]$$

Knowing that the *entire* shaft surface is selected, but only that portion of the shaft surface within the contact region *should* be considered, the above difference is acceptable.

This concludes the analysis portion of the current example. The following section describes a semi-automatic method of generating a **Report** within COSMOSWorks.

Generating a Report

COSMOSWorks provides a pre-defined **Report** format for quickly generating and sharing information related to any Study. The **Report** folder contains a basic outline of items that the user can choose either to include or to exclude from a report. A report also provides a means for sharing information with others either via the internet (an HTML file) or in print form (a Word® file). The process for including or excluding items from a report is outlined below.

1. Before generating a report, adjust the model size and orientation on the screen to best display graphical characteristics of interest to the user. This is done because

Interference Fit Analysis

the orientation of all plots produced within the report is identical to the current screen image.

2. In the COSMOSWorks manager, right-click **Report**. From the pop-up menu select **Define…** and the **Report Options** window opens as seen in Fig. 26.

At the top-center of this window, adjacent to **Report style:**, it is possible to select between **Contemporary** (default), **Professional** and **Elegant** style reports. These options are accessible via a pull-down menu. Because only slight font and layout differences exist between these report formats, accept the default **Contemporary** style and proceed as follows.

Immediately beneath the **Report style:** pull-down menu are two tables. When first opening the **Report Options** window the left-hand table is typically empty and the right-hand table contains a list of all items that can be included in a report.

To remove an item from a report the item is highlighted in the table (by clicking to select it) and then pressing the [<] button to move the selected item into the left-hand table. To move *all* items from the right table to the left table the [<<] button is pressed. Conversely, if one or all items are to be moved from the left-hand table into the right-hand table the [>] and [>>] buttons are used respectively. *Try the above.*

The order of items within a report can also be changed by highlighting the item to be moved and clicking either the **[Move up]** or **[Move down]** buttons located to the right of the tables. Based on this brief introduction, we next proceed to generate a brief report. *Try this, then restore the list to its original order.*

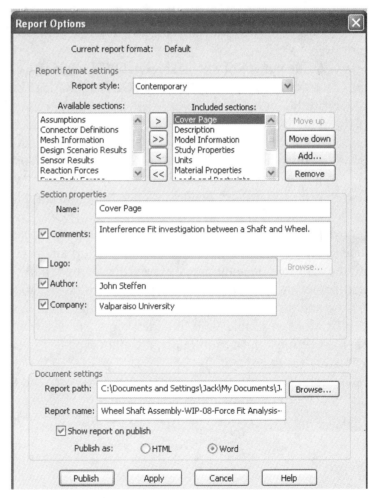

Figure 26 – Initial window of the **Report** folder showing a list of possible items that can be included or removed from a sample report. **The Cover Page** template is shown in the middle of the window.

In the report generated below, several items are removed for the sake of brevity.

3. Begin by highlighting the **Cover Page** (if not already selected). Then, check the ☑ **Author:** field and type your name. Next, check ☑ **Company:** and enter your company or university name. Finally, check ☑ **Comments:** and provide a brief description of the current problem. A report title is automatically assigned based on the Study name.

4. Next click to highlight the *word* **Description**. This action activates the **Description** section of the report. Check ☑ **Comments:** and type a description of the goals of this example in the text-box.

5. Proceed down the list of remaining items by highlighting each item and briefly viewing the information displayed and/or information you can add. The following list suggests sections to be *removed* [<] from this report because they either are not applicable or are not important to this Study. **Assumptions, Connector Definitions, Design Scenario Results, Sensor Results, Reaction Forces, Free-Body Forces, Bolt Forces,** and **Pin Forces**.

6. At the bottom of the list, highlight **Conclusion**. Check ☑ **Comments:** and in the text-box, type a brief set of conclusions that can be drawn from the results of this Study.

7. At the bottom of the **Report Options** window check the ☑ **Show report on publish** and in the **Publish as:** list choose ⦿ **Word**, then press the **[Publish]** button. These settings automatically generate a Word© version of the report. Typically a Word document containing the completed report opens after a few seconds.

Take several minutes to scroll through the report and review the information you entered and the information automatically recorded in the report folder. Notice the variety of details relevant to the current example that are saved. Items such as the mesh type, mesh quality, and mesh size, are included along with data about the FEA solver used, material specifications, restraints, and full color plots, etc. are *summarized* in the **Report**. While this file contains much useful data, it cannot replace the engineering insight that goes into evaluating results of a Study.

This concludes the present example. Unless instructed to save results of this example, proceed as follows to exit COSMOSWorks without saving the solution for this model.

1. Select **File** from the main menu at top of screen and from the pull-down menu select **Close**.

2. A **SolidWorks** window opens and prompts, "**Save changes to Wheel Shaft Assembly?**" Click **NO**.

EXERCISES

1. A bronze bushing is to be force fit into a hole in the cast iron wall of a machine frame. A bushing of this type serves as a bearing in a journal bearing set. The journal (i.e., the shaft) and bearing are separated by a thin film of lubricant, typically oil. Only the bushing and a portion of the machine frame are shown in Fig. E5-1; the shaft is not shown. Important dimensions of the bushing and wall are included. The wall thickness and bushing length are equal. It is assumed that a sufficiently large segment of the machine frame is included in the model below so that boundary effects, due to influences of St. Venant's principle, are insignificant. Create a finite element analysis of the interference fit between these two mating parts.
   Open the file: **Bushing and Frame 5-1**.

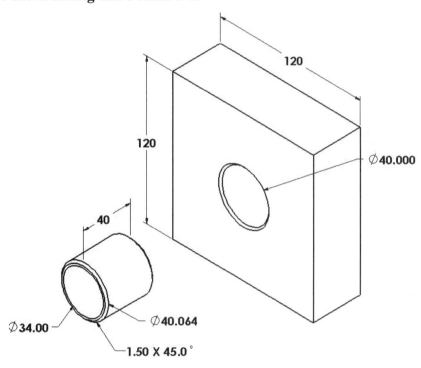

Figure E5-1 – Bushing and a portion of a machine frame wall. A comparison of the bushing outside diameter (o.d. = 40.064 mm) and the hole inside diameter (i.d. = 40.000 mm) in the machine frame reveals an interference fit between these mating parts.

- Material: Machine frame – **Gray Cast Iron** (Use S.I. units)
    Bushing – **Tin Bearing Bronze** (listed beneath "**Copper Alloys**")

- Mesh: **High Quality**, default size tetrahedral elements

- Restraint: Set **Contact/Gaps** to **Shrink Fit** between the bushing and machine frame. If necessary, add restraints to prevent rigid body motion.

Determine the following:

a. Develop a finite element model that includes: material specification, restraints, a meshed model and a solution. Defeature the model prior to analysis.

> **Solution Guidance**
> Discussion below provides general guidance (i.e., hints and reminders) for the solution to this problem. In particular instances where this solution differs from the example problem of this chapter, specific steps are provided.
>
> Materials for this example are not pre-specified in SolidWorks due to the more limited material selection found there. Therefore, after creating a **Study** in COSMOSWorks, assign *different* materials to each part as follows.
> - In the COSMOSWorks manager, click the "+" sign adjacent to the + **Solids** folder. The **Bushing-2** and **Frame-1** folders appear.
> - Right-click the **Bushing-2** folder to open the pull-down menu shown in Fig. E5-2.
> - From the pull-down menu, select **Apply Material to All Bodies...**
> - In the **Material** window, specify the bushing material in the usual manner.
> - Repeat the above steps to define **Gray Cast Iron** for the machine frame.
>
>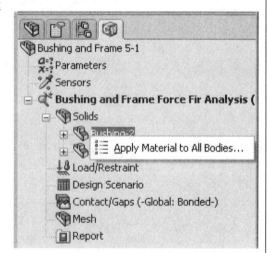
>
> Figure E5-2 – Menu selections used to specify *different* materials for different parts of an assembly.
>
> The above procedure applies the same material to all parts contained in each sub-folder. In this instance there is only a single part in each sub-folder. However, if for example, different bushings were made of different bronze materials or if parts of the machine frame were made from a material other than Gray Cast Iron, then it would be necessary to go one level deeper to assign different materials to different parts. This would be accomplished by clicking the "+" signs adjacent to **Bushing-2** and **Frame-1** respectively.
>
> Next defeature the model and reduce it to ¼ of the given assembly.
> - Toggle to SolidWorks and beneath + **Frame<1>**, **Suppress** the **Frame Chamfer**. Similarly, beneath **Bushing<2>**, **Suppress** the **Bushing Chamfer**
> - Also in the SolidWorks manager tree, right-click **Quarter Model – Assembly** and from the pop-up menu select the **Unsuppress** icon.
>
> When applying restraints to the model, consider carefully what restraint(s), if any, must be specified to prevent rigid-body-motion.

b. Perform an interference check between the bushing and machine frame. Plot the resulting image.

c. Using classical equations, compute circumferential and radial stresses at the inner surface of the machine frame and the outer surface of the bushing. It is common practice to use the *radius* of a circle tangent to the inside of the machine frame as the outside "radius" of the outer member. Also determine the common contact pressure between these surfaces.

d. Create a stress contour plot of von Mises stress in the assembly. Include automatic labeling of maximum and minimum von Mises stress on this plot.

e. In a cylindrical coordinate system, plot and compare circumferential stress at the inner surface of the machine frame with results calculated using classical equations determined in part (c). The value to be used for the finite element stress on the frame is the **Avg.** value of circumferential stress determined by using the **List Selected** feature applied to the inner surface of the hole in the frame. If necessary, click-and-drag to adjust the column width to show all digits of tabulated data. Compute the percent difference between classical and finite element analysis results using equation [1].

$$\% \text{ difference} = \frac{(\text{FEA result - classical result})}{\text{FEA result}} * 100 = \qquad [1]$$

f. Repeat part (e) for circumferential stress at the outer surface of the bushing.

g. In a cylindrical coordinate system, plot and determine the average value of radial stress (contact pressure) between mating surfaces using the **List Selected** feature. Apply the **List Selected** feature to the inner surface of the hole and outer surface of the bushing. Although, theoretically, these values should be the same, some differences may occur. Therefore, determine the average of these two values and compare it with the classical result predicted for contact pressure. Then determine the percent difference by again using equation [1].

h. Because length of inner and outer members is equal, determine the radial displacement of the bushing (inner member) and machine frame (outer member) using the average **(Avg.)** value determined by the **List Selected** option. Determine the total radial interference and compare it with the given interference. Comment upon whether or not they check, and if not, explain why not?

Textbook Problems
In addition to the above exercise, it is highly recommended that additional problems involving interference fits between mating parts be worked from a design of machine elements textbook. Textbook problems provide a great way to discover errors made in formulating a finite element analysis because they typically are well defined problems for which the solution is known. Typical textbook problems, if well defined in advance, make an excellent source of solutions for comparison.

CHAPTER #6

CONTACT ANALYSIS

Examples up to this point have included models for which various preliminary steps, such as suppressing or un-suppressing certain model features, were pre-planned so that examples could proceed quickly to introduce new principles and techniques. However, a real analysis is not often delivered in such a "ready to go" condition unless the user, or a CAD technician, plans ahead to prepare a model in the desired format. Either case requires the finite element analyst either to develop the model according to their own expectations or to inform a CAD expert about special modeling needs. To better prepare FEA users for these more realistic scenarios, this example includes many of the necessary preparatory steps thereby creating a more realistic example from beginning to end.

Learning Objectives
Upon completion of this example, users should be able to:

- Assign loads to a *non-flat surface* in *specified direction*(s)

- Define *Contact/Gap* conditions with *no penetration*

- Use *Animation* to understand deformation and stress development as load(s) are applied

- Apply *Iso Clipping* to view model stresses

- Display *Contact Pressure* plots

Problem Statement

A trunion mount, of the type used to attach hydraulic or pneumatic cylinders to a fixed surface, is shown in Fig. 1. Both the trunion and pin are made of **Alloy Steel**. The pin is subject to an 800 lb force acting upward to the right at $60°$ from the horizontal as shown in the right-side view. The goal of this analysis is to determine maximum von Mises stress and its location on the trunion mount and to determine contact pressure distribution between the pin and holes in the trunion.

Contacting surfaces occur where the pin passes through holes in the side-plates of the trunion and where the mid-portion of the pin is acted upon by a cylinder force (the cylinder attachment is not shown). For this reason, it is necessary to create separately identifiable contacting surfaces on the pin. *Split Lines* are used to divide the pin surface.

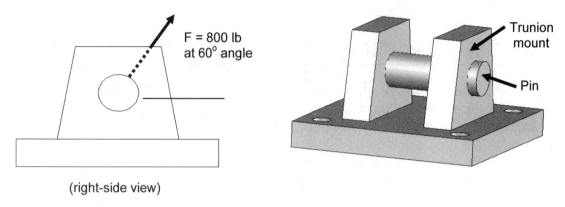

Figure 1 – Right-side view of trunion mount and pin showing direction of applied force.

The procedure outlined below is not the only way to apply *Split Lines* to the pin. It is, however, deemed simpler to demonstrate this aspect of model preparation on the pin alone, before joining it with the trunion base to form an assembly. Proceed as follows. *(Because a personal goal should be to develop independent competence using COSMOSWorks, many steps below are abbreviated where prior experience should be sufficient).*

Preparing the Model for Analysis

1. Open **SolidWorks 2008**.

2. Select **Files / Open** and open the part file named "**Trunion Pin**."

All steps related to locating and creating *Split Lines* occur within SolidWorks, therefore do not open COSMOSWorks at this time. Three *Split Lines* are to be added to the trunion pin at locations illustrated on Fig. 2. It is important to know that the origin of the coordinate system is located at the center of the pin and that the pin is centered in the trunion mount. Given this information, those who are comfortable in their ability to locate the necessary reference planes and *Split Lines* are encouraged to proceed on their own. These individuals can skip the next two sections, titled "Add Reference Planes" and "Insert Split Lines." However, all necessary steps are provided below if guidance is desired.

Split Line locations

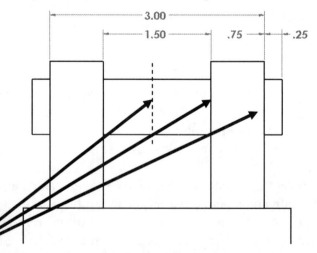

Figure 2 – Location of *Split Lines* at the intersection of the pin and side supports of the trunion mount and at the center of the pin.

Add Reference Planes

1. From the main menu select **Insert**.

2. In the pull-down menu, highlight **Reference Geometry** and from the subsequent pop-up menu, select **Plane...** The **Plane** property manager opens and the SolidWorks flyout menu appears as seen in Fig. 3. If the flyout menu is not open, click the "+" **Trunion Pin** icon at top of the flyout menu to open it.

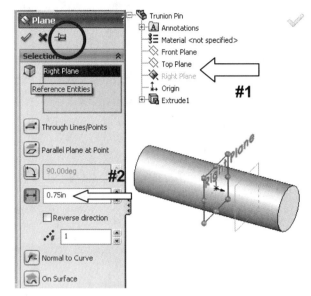

Figure 3 – Selecting the **Right Plane** as a starting point, the first **Reference Plane** is located 0.75 in to its right.

3. Click the **Keep Visible** "push-pin" at top of the **Plane** property manager (circled on Fig. 3). This action keeps the **Plane** property manager open so that multiple steps can be performed without the need to re-open it for each change.

4. In the **Selections** dialogue box the **Reference Entities** field is active (light blue). In the SolidWorks flyout menu, select **Right Plane**. See arrow #1 in Fig. 3.

5. In the **Offset Distance** spin box, at arrow #2, type **0.75** as the distance from the **Right Plane** to the location of the first **Reference Plane**. An outline of the **Reference Plane** appears on the pin.

6. Click **[OK]** ✓. **Plane1** is listed at the bottom of the SolidWorks flyout menu and is labeled on the pin.

7. Because the **Keep Visible** push-pin is active, the **Reference Entities** field is updated and displays **Plane1**. Also, the previous value (**0.75**) entered into the **Offset Distance** spin box remains unchanged.

By coincidence, the next reference plane is located at its desired position 0.75 in to the right of **Plane1** and is displayed on the model. Click **[OK]** ✓ to accept this new reference plane and **Plane2** is also added to the bottom of the SolidWorks flyout menu and is labeled on the pin.

8. Click to deselect the **Keep Visible** push-pin.

9. Repeat steps 4 to 6 again, but this time type **0.0** as the distance from the **Right Plane**. In other words, the third reference plane is added *at* the location of the

Right Plane. A plane is located at the pin center because symmetry is used later in this example and both the trunion mount and the pin are cut at this location.

10. Click **[OK]** ✓ to close the **Plane** property manager.

A model of the pin and its three **Reference Planes** should appear as shown in Fig. 4.

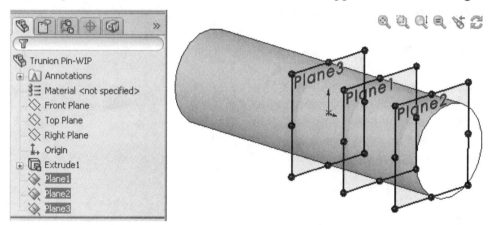

Figure 4 – Three **Reference Planes** are created to denote locations where *Split Lines* are to be added to the part.

Insert Split Lines

Intersections of **Plane1**, **Plane2** and **Plane3** with the pin are used to locate *Split Lines* on the pin as outlined below.

1. In the Main menu select **Insert**.

2. From the pull-down menu highlight **Curve** and from the pop-up menu, select **Split Line…**. The **Split Line** property manager opens as shown in Fig. 5.

3. Because a **Split Line** is to be located where each plane intersects the pin, in the **Type of Split** dialogue box, choose ⦿ **Intersection**.

4. In the upper field of the **Selections** dialogue box, the **Splitting Bodies/Faces/Planes** field is highlighted. Either select each of the planes appearing on the pin, or click **Plane1**, **Plane2** and **Plane3** in the SolidWorks flyout menu. Names of the three planes are listed in the upper field in the order selected.

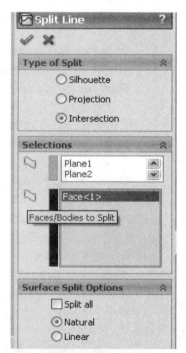

Figure 5 – The **Split Line** property manager is used to determine locations of *Split Lines* on the model.

5. Next, move the cursor onto the second field; its name appears as **Faces/Bodies to Split**. Click to activate this field. Then, move the cursor into the graphics area and select the *cylindrical* surface of the pin. **Face<1>** appears in this field.

6. In the **Surface Split Options** dialogue box, the setting should appear as ⊙ **Natural**.

7. Click **[OK]** ✓ and three *Split Lines* appear on the pin as shown in Fig. 6. Also, **Split Line1** is listed at the bottom of the SolidWorks manager tree.

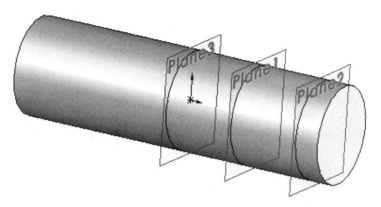

Figure 6 – *Split Lines* located on the trunion pin surface at intersections with the three **Reference Planes**.

Because symmetry exists, only the right-half of the pin is used for analysis. Therefore, no additional *Split Lines* need be added to the other half of the model. Save this file as outlined below.

8. From the main menu, select **File** followed by **Save As…**.

9. In the **Save As** window, name the part "**Pin with Split Lines**" and **[Save]** the file to a PC, to network file space, or to temporary storage media such as a jump drive.

10. Close the file by selecting **File / Close**. (Do *not* close COSMOSWorks.)

Creating the Assembly Model

It is assumed that most COSMOSWorks users are already familiar with the SolidWorks work environment and the process of creating an *assembly* composed of two or more *parts*. Those individuals can proceed to join the **Trunion Base** and the **Pin with Split Lines** to form an assembly on your own and skip to the section titled **Create a Finite Element Analysis (Study)**. However, for many users who are new to both COSMOSWorks and SolidWorks, a step-by-step procedure to create the assembly is outlined below.

In the assembly the pin is located concentric with holes in the trunion base and extends 1/4 inch outside either of the two side supports as shown in Fig. 2. This location will be replicated in the assembly created in the following steps.

1. In the main menu, click **File / New....** The **New SolidWorks Document** window opens.

2. Click the **Assembly** icon and then click **[OK]**. The **Begin Assembly** property manager opens as shown in Fig. 7.

3. Click to activate the **Keep Visible** push-pin circled at the top-right of the property manager and read the message highlighted in yellow.

4. Select the **[Browse...]** button at bottom of the **Part/Assembly to Insert** dialogue box and the **Open** window appears. This action should take the user to the location where the **Trunion Base** and **Pin with Split Lines** are stored. If not, browse to other file storage locations as needed.

5. From the list of files, select **Trunion Base** and click **[Open]**.

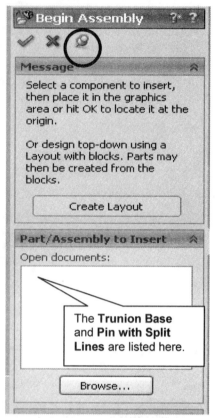

Figure 7 – The **Begin Assembly** property manager used to bring *parts* into an *assembly*.

6. The cursor and part appear and move together in the graphics area. Move the cursor to the top of the **Begin Assembly** property manager and click **[OK]** ✓. The trunion base automatically moves to the coordinate system origin. If the cursor is moved into the graphics area a second trunion base appears on the screen. *Do not click again!*

Ignore the second trunion base and proceed directly to the next step.

7. Move the cursor into the **Begin Assembly** property manager and again select the **[Browse...]** button. The **Open** window appears.

8. In the **Open** window, select the **Pin with Split Lines** and click **[Open]**. The cursor and pin appear and move together on the graphics screen.

9. Move the pin to a position above and to the left of the trunion base as illustrated in Fig. 8 and click to place it there. A small image of a mouse appears with a green check-mark "✓" on the right-mouse button, click the right mouse button to accept this position. Finally, click **[OK]** to close the **Begin Assembly** property manager. *(The actual pin location is arbitrary, but first time users will gain*

greater insight into the assembly process if the pin and base do not initially intersect).

The next task is to assemble the pin into holes on the trunion base. This requires use of a **Mate** definition between the parts to be joined. Because we are currently working in the *assembly* mode, the **Assembly** tool bar, shown in Fig. 9, should be displayed somewhere on the screen. If not, several alternate methods of accessing the **Assembly** toolbar are available, two are outlined below.

Method 1

- Right-click anywhere in the toolbar at top of the screen. The **Command Manager** menu opens.

- From this menu select **Assembly** and the toolbar shown in Fig. 9 should appear.

- Skip to the "CAUTION" below Fig. 9.

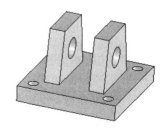

Figure 8 – Trunion base and pin (in an arbitrary position) prior to assembly.

Method 2

- In the main menu, select **View**.

- From the pop-up menu select **Toolbars** ▶.

- From the next pop-up menu select **Assembly** and the toolbar in Fig. 9 should appear.

- Skip to the "CAUTION" below Fig. 9.

Figure 9 – **Assembly** toolbar with the **Mates** icon indicated. The **Mates** property manager is used to define geometric relationships (i.e., locations) of one part relative to another.

CAUTION: In the steps that follow the **Mate** icon is a single paper clip; it is *not* to be confused with the two paperclips labeled **Mates** located at the bottom of the SolidWorks feature manager tree.

10. Click the **Mate** icon .

11. The **Mate** property manager opens as seen in Fig. 10. The **Mate Selections** dialogue box is active (highlighted light blue). Passing the cursor over this field indicates it is to be filled with the **Entities to Mate**.

The first step is to position the pin so that it is concentric (aligned) with holes in the trunion base. To accomplish this, the two parts are selected as described next. NOTE: Because *Split Lines* are defined on the pin, its surface is effectively subdivided into segments at each *Split Line*. However, as described in the following step it is necessary to select *only one* cylindrical segment on the pin.

12. Click to select *any* cylindrical surface on the pin and the *cylindrical* surface *inside* either hole as highlighted in Fig. 11. The two parts move to align their cylindrical surfaces and names of the two selected faces appear in the **Mate Selections** field. The software is "smart" enough to guess that a concentric mate is probably desired, but this is not confirmed until the next step.

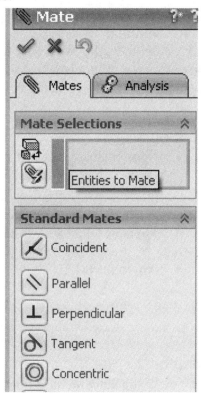

Figure 10 – Initial selection of surfaces to be specified as concentric is made in the **Mate** property manager.

A small pop-up icon bar appears as shown in Fig. 11. It lists icons of possible mates that might be defined between the two cylindrical surfaces.

13. It is suggested that new users select the **Concentric** icon mate in the **Mate** property manager because its name is listed adjacent to its icon. However, both icons are circled in Fig. 11 and either can be selected.

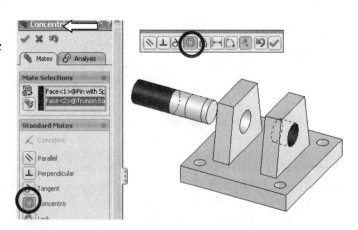

Figure 11 – Defining a **Concentric** mate between the pin and trunion base holes results in alignment between the two parts.

Contact Analysis

After selecting **Concentric**, the name **Concentric1** appears at top of the property manager (see arrow in Fig. 11) and names of the two faces are listed in the **Entities to Mate** field.

14. Click **[OK]**✓ to close the **Concentric1** property manager. *(Do NOT click [OK] twice; see next step).*

The property manager remains open, but its name changes back to the **Mate**. Proceed as follows to further define pin location within the holes

15. Click the right *end* of the pin *and* the right-face of the trunion support shown highlighted in Fig. 12. The pin-end is located flush with the trunion face.

However, because the 3.5 in long pin is to be centered between the two supports (whose outside dimension is 3.00 in, see Fig. 2), pin location is altered as follows.

16. Type **0.25** inches into the **Distance** spin box indicated by an arrow on Fig. 12. This value ensures that 0.25 inch of the pin is offset from the two selected surfaces. *See next step.*

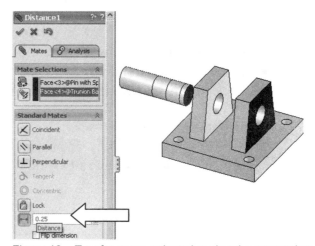

Figure 12 – Two faces are selected so that the appropriate **Distance** between them can be defined to center the pin between holes on the trunion base.

17. If the pin appears recessed within the right-side hole, it may be necessary to check ☑ **Flip Direction** located below the **Distance** spin box. The assembly should now appear as shown in Fig. 1, repeated below.

18. Click **[OK]** *twice* to close the **Distance1** and then the **Mate** property managers.

19. To protect the work invested in this example thus far, from the main menu select **File / Save As…** and in the **Save As** window, name the assembly "**Trunion and Pin Assembly**" then click **[Save]** to save the file.

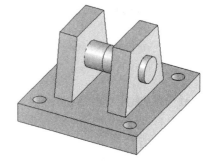

Figure 1 – Repeated

This concludes the somewhat lengthy process of properly preparing a model for finite element analysis. Although these, or similar steps, were omitted from previous problems, one goal of the current example is to develop a comprehensive picture of typical tasks involved in formulating a complete analysis. The following sections define and solve the finite element model.

Create a Finite Element Analysis (Study)

1. Toggle to the **COSMOSWorks** analysis manager.

2. Right-click **Trunion and Pin Assembly** and from the pull-down menu select **Study…**

3. In the **Name** field, type **Trunion and Pin Contact Analysis** to identify this study.

4. Verify that a **Solid mesh** and a **Static** analysis are selected then click **[OK]** ✓ to close the **Study** property manager.

5. Right-click **Trunion and Pin Assembly** and from the pull-down menu select **Options…**. The **Systems Options - General** window opens.

6. Under the **Default Options** tab, select **Units** and set the following options:
   - In the right-half of the window under **Unit system**, select ⊙ **English (IPS)**
   - Under **Units** set **Length/Displacement:** to **in** and **Stress/Pressure:** to **psi**.

   Ignore remaining options.

7. Click **[OK]** to close the **Default Options - Units** window.

Assign Material Properties

1. Right-click the **Solids** folder and select, **Apply Material to All…**.

2. In the **Material** window, select ⊙ **From library files** and verify that **cosmos materials** appears in the pull-down menu.

3. From the list of available steels, select **Alloy Steel (SS)**. In the right-half of the **Material** window, notice that **Units:** appear as **English (IPS)** as defined above.

4. Click **[OK]** to close the **Material** window.

Cut Model on Symmetry Plane

This example makes use of model symmetry to realize computational efficiencies associated with a smaller model and to provide additional practice in using this powerful software feature. Steps below lead the user through one of several possible procedures that can be used to create a symmetrical half-model.

1. Toggle to the **SolidWorks** manager tree.

2. In the toolbar, click the **Front** view icon to rotate the front plane of the model into the plane of the screen.

3. Next, in the SolidWorks manager tree, click to select **Front Plane**; see arrow in Fig. 13. The **Front Plane** is shown on an image of the model in Fig. 13.

NOTE: Additional top, front and right planes exist for both the trunion base and for the pin. Although consistency of plane location was carefully coordinated for this example, it is important to realize that individual *parts* are created in their own local coordinate systems. These individual local coordinate systems may be oriented differently once parts are brought together in the global coordinate system of an *assembly*.

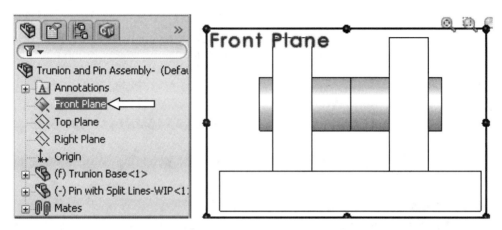

Figure 13 – The **Front Plane** is selected as the plane to sketch on as part of the process of cutting the model in half.

4. Click the **Sketch** icon to open a sketch on the front plane. If this toolbar does not appear on the screen, from the main menu, select **View**. From the pull-down menu, select **Toolbars▶**. Then, scroll down in the toolbars menu and select the **Sketch** icon. This action opens the **Sketch** toolbar.

5. In the **Sketch** toolbar, select the **Line** icon and move the cursor onto the graphics screen. The cursor changes to a pencil symbol with a line adjacent to it.

6. Move the cursor above (or below) the *middle* of the assembly and a vertical dashed line appears to indicate alignment with the origin of the coordinate system located there. At his location, click and move the mouse to create a vertical line that extends completely through the model and extend it beyond its opposite border, then release the mouse button. To end the line, click again.

7. Press the **[Esc]** key to terminate line drawing. A screen image of the assembly should look like the right-half of Fig. 14.

Aside:
Recall when the trunion base was brought into the assembly drawing that it was positioned at the "center" of the assembly coordinate system.

Analysis of Machine Elements using COSMOSWorks

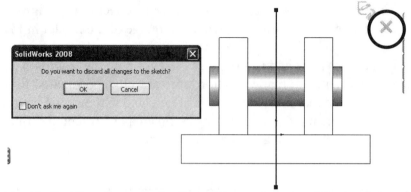

Figure 14 – Vertical line drawn on the **Front Plane** at center of the trunion and pin assembly. Also shown is the warning window that appears when closing a sketch.

8. Click the red "X" at upper right corner of the screen, circled in Fig. 14, to close the sketch. The **SolidWorks** warning window opens as shown on the left-side of Fig. 14 and prompts, "**Do you want to discard all changes to the sketch?**"

9. Click the **[Cancel]** button because it is desired to keep the changes made above.

10. Select the **Extrude Cut** icon and the **Cut-Extrude** property manager opens. A portion of this property manager is illustrated in Fig. 15. *NOTE: If, instead, the Extrude property manager opens, click to select the vertical line again. Doing so opens the Cut-Extrude property manager.*

11. The **From** dialogue box should indicate **Sketch Plane** and the **Direction1** dialogue box should indicate **Through All**.

12. If necessary, rotate the model so the surface created by extending the vertical line into a cutting plane is visible slicing through the model as shown in Fig. 16.

Three arrows appear on the plane. They point in directions in which the model is cut and material removed. Carefully observe the direction of the arrow *normal* to the plane. It should point *away* from the right-half of the model. Recall that the right-half of the pin was modified by the addition of *Split Lines*. Thus, the right-half of the model is to be saved.

13. If the normal arrow is directed toward the right-half of the model, click to change its direction by selecting ☑ **Flip side to cut**. When the normal arrow is directed in the correct direction, toward the left-side, proceed to the next step.

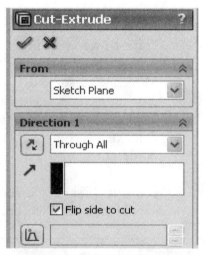

Figure 15 – **Cut-Extrude** property manager used to define direction for portion of the assembly that is cut and removed by the sketch plane.

6-12

Contact Analysis

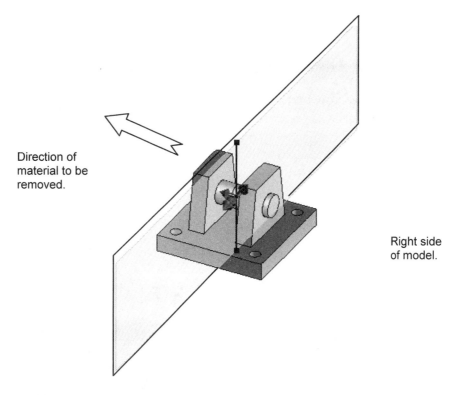

Figure 16 – The cutting plane created by extending the vertical line into a plane is shown along with arrows indicating the direction in which material is to be removed.

14. Click **[OK]** ✓ to close the **Cut-Extrude** property manager. The assembly should now appear as shown in Fig. 17.

15. Toggle back to COSMOSWorks by selecting the COSMOSWorks icon at top of the manager tree.

Next, several types of restraints, including: **Immovable**, **Symmetry** and **Contact/Gaps**, are applied to the model. All these restraint types were used before. However, a different aspect of the **Contact/Gaps** restraint is used in this example.

The following section includes subtitles to identify specific restraint types applied to the model. Thus, individuals who choose to skip a section and apply restraints on their own are encouraged to do so.

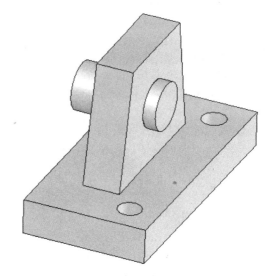

Figure 17 – Assembly after cutting it in half to make use of geometric symmetry.

6-13

Assign Restraints and Loads

Symmetry and Immovable Restraints

In the following steps, **Immovable** and **Symmetry** restraints are applied to bolt holes in the trunion base and to cut-surfaces respectively. An abbreviated outline of necessary steps is provided below.

1. Right-click the **Load/Restraint** folder and from the pull-down menu, select **Restraints…**.

2. In the **Type** dialogue box, select **Immovable (No translation)** from the pull-down menu.

3. Move the cursor over the model and zoom in, as necessary, to select the *inside surface* of each bolt hole on the trunion base. Restraint symbols appear inside each hole as illustrated in Fig. 18.

4. Click **[OK]**✓ to close the **Restraint** property manager.

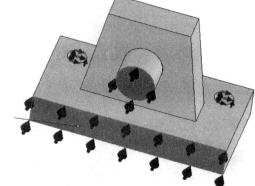

Figure 18 – **Immovable** restraints applied to bolt holes and **Symmetry** restraints applied to cut surfaces of the pin and trunion base.

5. Again, right-click the **Load/Restraint** folder and from the pull-down menu, select **Restraints…**.

6. In the **Type** dialogue box, select **Symmetry** from the pull-down menu. To apply this restraint select both cut surfaces highlighted on the model in Fig. 18. The **Symmetry** restraint is applied normal to the cut surface, thereby preventing displacement in the X-direction. Compare directions with the coordinate system triad to verify this statement. NOTE: Restraint symbols are typically directed toward the cut surfaces, however, their direction makes no difference.

7. Click **[OK]** ✓ to close the **Restraint** property manager.

Contact/Gaps Restraints

In the following steps, **Contact/Gaps** restraints are used to define local contact between the pin and inner surface of the hole in the trunion base. An exploded view of the assembly is created to facilitate selection of the contacting surfaces. Proceed as follows.

1. In the main menu click **Insert** and from the pull-down menu select **Exploded View…**. The **Explode** property manager opens with the message, "**Select component(s) and then drag the manipulator handle to create an explode step.**"

Contact Analysis

2. Click to select the pin. Then *drag* the manipulator arrow in the direction of the X-axis (this direction corresponds to the pin axis) to move the pin away from the model as shown in Fig. 19.

3. Click **[OK]** ✓ to close the **Explode** property manager.

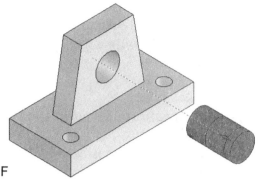
F
Figure 19 – Explode the assembly to facilitate selection of contacting surfaces.

4. In the COSMOSWorks manager tree, right-click the **Contact/Gaps (-Global Bonded-)** folder and from the pull-down menu, select **Define Contact Set...**. The **Contact Set** property manager opens as shown in Fig. 20.

5. In the **Type** dialogue box, select **No penetration** as the type of contact defined between the pin and hole surfaces.

6. Select the *inner surface* of the hole, highlighted in Fig. 20, and **Face<1>@ Trunion Base-1** is listed in the **Faces, Edges, Vertices for Source** in the upper field of the **Type:** dialogue box.

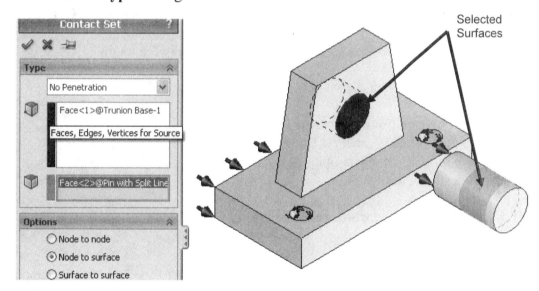
Figure 20 – Selecting surfaces between which the **Contact/Gaps** restraint is applied.

7. Next, click the **Faces for Target** field (second field from top) to activate it. Then, on the pin, select the corresponding surface, also highlighted in Fig. 20. **Face<2> @Pin with Split Line** appears in the **Faces for Target** field.

8. In the **Options** dialogue box, select ⊙ **Node to surface**.

9. Click **[OK]** ✓ to close the **Contact Set** property manager. Highlighting disappears from the model and beneath the **Contact/Gaps** folder appears an icon labeled, "**Contact Set-1 (-No Penetration<Trunion Base-1, Pin with Split Lines-1>-).**"

Apply a Directional Load to the Pin

This section outlines steps necessary to apply a directional load to the pin. In previous examples, only force components in the X, Y and Z directions were applied to parts. Using that approach, the force applied to the pin at a 60° angle from the horizontal (Fig. 1) would be represented by the Y and Z-components of force on the model. However, this example demonstrates application of these force components in a way not used in earlier problems. Further, because only half of the model is being analyzed, only half of the entire load is applied to the pin. However, before applying the load, the model is re-assembled as follows.

1. Toggle to the **Configuration** manager icon located just to the left of the COSMOSWorks manager icon.

2. At top of the **Configuration manager**, click the "+" sign adjacent to **Default <Display State-1> [Trunion Pin and Assembly]** to display additional folders.

3. Next, right-click **ExplView1** and from the pull-down menu, select **Collapse**.

After the model is re-assembled, proceed as follows to apply the load.

4. Toggle back to the COSMOSWorks manager.

5. Right-click the **Load/Restraint** folder, and from the pull-down menu select **Force…**. The **Force** property manager opens; *only* its upper-half is shown in Fig. 21.

6. In the **Type** dialogue box, select ⦿ **Apply force/moment**.

7. The **Faces, Edges, Vertices, Reference Points for Force** field is highlighted and awaits selection of the entity to which the force is applied. Rotate the model as necessary and click the pin surface shown highlighted in Fig. 22. **<Face1>@Pin with Split Lines** is listed in the active field.

Figure 21 – Selecting faces to define force component direction in the **Force** property manager.

6-16

8. In the **Type** dialogue box, click to activate the **Face, Edge, Plane, Axis for Direction** field (second field from top) and proceed to select the trunion support face highlighted in Fig. 22. **Face<2>@Trunion Base-1** appears in the active field.

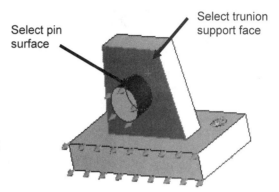

Aside: It is also possible to select any other face, edge, plane or axis that is aligned with the direction of the applied force components. Other surfaces that could be selected include: the cut-end of the pin; the cut-face of the trunion base; or the **Right Plane** from the SolidWorks flyout menu (the plane selected *must* be at the assembly level, *not* at the component level).

Figure 22 – Selecting the surface for force application and a face to define its direction.

The complete **Force** property manager is shown in Fig. 23. In the **Force (Per entity)** dialogue box, three choices are available for defining magnitude and direction of a force. Icons adjacent to the top two fields show two different directions for applying forces parallel to the selected face while the bottom field indicates a force applied perpendicular to the selected face.

9. In the **Force (Per entity)** dialogue box, move the cursor over the top field to reveal its label as **Along Plane Dir 1**. Click the icon adjacent to this field to activate it. Immediately, force vectors are displayed on the center segment of the pin. (*Ignore their direction for now.*)

10. These vectors are parallel to the trunion face selected above. Use the coordinate system triad to verify that these vectors lie in the Z-direction. This direction corresponds to the horizontal component of the 400 lb force[1]. Therefore, in the top field, type **200**, which is determined from:

$$F_Z = 400 * \cos 60° = 200 \text{ lb}$$

Figure 1 (repeated below) shows geometry related to the above calculation. Also observe on-screen "flags" that display direction, units and magnitude of force components.

Figure 23 – Specifying magnitude and direction of force components.

---

[1] Recall, half the total force of 800 pounds is applied to the half-model.

11. If force vectors are oriented in the incorrect direction, check ☑ **Reverse direction**. Refer to Fig. 24 for proper directions of force components.

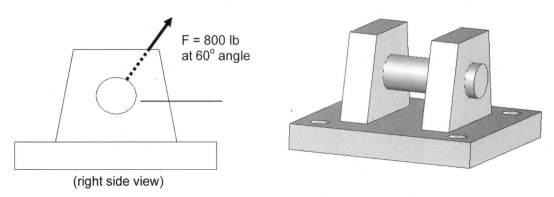

(right side view)

Figure 1 (repeated) – Right-side view of trunion mount and pin showing direction of applied force.

12. Enter the Y-component of force on your own and verify that its direction is upward as shown on Fig. 24. Use a force magnitude of:

$$F_Y = 400*\sin 60° = 346.4 \text{ lb}$$

13. Make certain the force **Normal to plane** is deselected (i.e., the field is "grayed" out).

14. Click **[OK]** ✓ to close the **Force** property manager

Y and Z force components applied to the pin should now appear as shown in Fig. 24. **Symmetry** and **Immovable** restraints are temporarily hidden to reduce clutter.

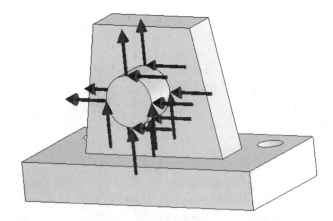

Figure 24 – To improve visibility, the Y and Z components of force applied to the pin are illustrated to a somewhat enlarged scale.

At this point the assembly is ready for meshing as outlined next.

Meshing the Model and Running the Solution

1. Right-click the **Mesh** folder and from the pull-down menu select **Create Mesh…**.

2. In the **Mesh** property manager check ☑ **Run analysis after meshing**. Also, accept the default mesh settings and click **[OK]** ✓ to close the **Mesh** property manager.

Notice that the solution takes longer to run (approximately 30 seconds on a Pentium 4 PC) especially during the **Solving contact constraints:** portion of the analysis.

Results Analysis

A primary goal of this example is to introduce tools used to observe contact pressure (stress) between mating components. As such, and because the state of stress is too complex to model with simple classical stress equations, little attention is focused on other stress results in the current model. However, because von Mises stress is widely used to compare its magnitude against material Yield Strength, its results are examined briefly below.

Von Mises Stress

1. Below the **Results** folder, double-click **Stress1 (-vonMises-)**. An image of von Mises stress distribution in the model is displayed in Fig. 25 and on your screen.

2. If the model does not appear as a deformed view, right-click **Stress1 (-vonMises-)** and select **Edit Definition…** Within the **Stress Plot** property manager, verify that a check appears adjacent to ☑ **Deformed Shape** and select ⊙ **Automatic** as the scaling method.

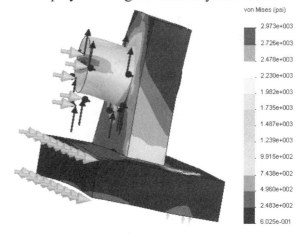

Figure 25– von Mises stress contours displayed on the deformed shape of the trunion and pin assembly.

3. Click **[OK]** ✓ to close the **Stress Plot** property manager. The displayed image should appear similar to that shown in Fig. 25.

Observe that the maximum von Mises stress (2,973 psi) is well below the material Yield Strength (89,984 psi). Recall that Yield Strength is *not* displayed on assembly plots because of the possibility that dissimilar materials might be used for different components in an assembly.

Iso Clipping

Unlike **Section Clipping** that permits stepping through a model at user specified increments of *distance* to view stresses on different "slices" of the model, **Iso Clipping** permits a variety of stress display options based on *stress level* to aid interpretation of study results. These options are investigated below.

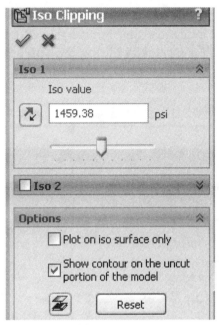

1. Right-click **Stress1 (-vonMises-)** and from the pull-down menu select **Iso Clipping…**. A partial view of the **Iso Clipping** property manager is illustrated in Fig. 26.

The **Iso 1** dialogue box contains a sliding scale whose two extreme values correspond to the minimum and maximum values of the quantity displayed at top of the color-coded stress legend appearing on the graphics screen. In the present case, von Mises stress magnitudes are currently displayed.

Figure 26 – Sliding stress scale in the **Iso Clipping** property manager.

2. In the **Iso 1** box, click-and-*drag* the sliding scale pointer and observe stress levels change on the model. Simultaneously, a moving arrow, adjacent to the color-coded stress legend, indicates the current stress magnitude being displayed on the image. Sliding the pointer from left-to-right results in lower stress levels being peeled away such that only areas of high stress remain. The current value of stress magnitude also appears in the **Iso value** box in the **Iso 1** dialogue box.

3. By clicking the **Reverse clipping direction** icon and then moving the sliding scale pointer from left-to-right, gradually increases the stress levels displayed on the model.

4. Next, in the **Options** dialogue box, located at bottom of the **Iso Clipping** property manager in Fig. 27, check ☑ **Plot on iso surface only** and once again move the sliding scale pointer. This time observe that *only* stress of a certain magnitude is plotted. This result is *not* pictured on the model in Fig. 27.

This option permits easy identification of regions where stress is *at* a certain level.

5. Clear the check mark from ☐ **Plot on Iso surface only**, before proceeding.

Contact Analysis

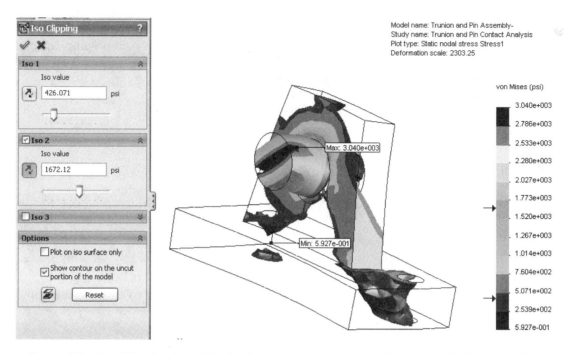

Figure 27 – **Iso Clipping** is set to display stresses *between* values set in the **Iso 1** and **Iso 2** dialogue boxes.

6. Next, check ☑ **Iso 2** and a second dialogue box opens in the **Iso Clipping** property manager shown at the left of Fig. 27.

7. Use the sliding scale in each dialogue box to set a lower bound and an upper bound for stress magnitudes to be displayed. In Fig. 27 **Iso 1** is set at **426** psi and **Iso 2** is set at **1672** psi; approximations show whole numbers only.

The corresponding plot, shown at the right side of Fig. 27, displays stress values *between* approximately 426 psi to 1,672 psi. Also observe two arrows adjacent to the color-coded stress legend. These arrows mark the lower and upper bounds of stress magnitudes currently displayed. NOTE: It may be necessary to click the **Reverse clipping direction** icons to properly set upper and lower bounds.

Analysis Insight

Iso Clipping permits easy identification of regions where stress levels are *above or below* a given value.

Iso Clipping also permits isolating and displaying stresses *between* certain lower and upper limits.

Before proceeding, return the full display of von Mises stress contours to the model as follows.

8. Clear the check-mark "✓" from the ☐ **Iso 2** dialogue box.

9. In the **Iso 1** dialogue box, slide the scale pointer to the extreme left position to display the full range of stress contours on the model. If they do not appear, click the **Reverse clipping direction** icon.

10. Click **[OK]** ✓ to close the **Iso Clipping** property manager.

Animating Stress Results

COSMOSWorks **Animation** capability permits dynamic viewing of a model due to applied loads provided ☑ **Deformed Shape** was checked, as specified earlier in the "**von Mises Stress**" section. During animation the model is cycled from no load to maximum load while stress, displacement, or strain is displayed. Insight gained by viewing these variations can be valuable in determining whether or not the model is behaving as expected based on applied loads and restraints. Proceed as follows to animate the von Mises stress results.

1. Right-click **Stress1 (-vonMises-)** and from the pull-down menu select **Animate…**. The **Animation** property manager opens as shown in Fig. 28.

2. If the model is animated, click the **Stop** icon at top of the **Basics** dialogue box.

3. Within the **Basics** dialogue box, the top field controls the number of **Frames** (i.e., the number of still images that are played back in sequence to simulate continuous motion. Set the **Frames** spin-box value to **10**. This value is a user preference. Higher values create smoother, but slower, animations.

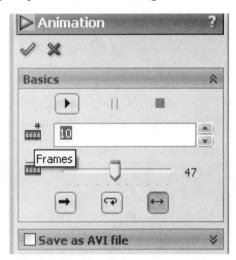

Figure 28 – A portion of the **Animation** property manager showing the **Frames** and **Speed** controls.

4. The slide-scale, located at the bottom of the **Basics** dialogue box, controls the **Speed** of animation. Moving the slide to the left slows the animation and, conversely, movement to the right speeds-up the animation.

Experiment with these capabilities on your own. The model can be rotated to view it from different angles while animation is proceeding. **Start**, **Pause**, and **Stop** buttons are located from left to right across the top of the **Animation** dialogue box. If desired, an animated sequence can be saved as an AVI File.

5. After experimenting with this capability, click **[OK]** ✓ to close the **Animation** property manager.

Animation of stress variation in the trunion mount corresponds to each power stroke of a cylinder. The cycling of stress from minimum to maximum should emphasize the need for fatigue analysis of many machine components.

Displacement Results

Displacement results can be displayed and examined in like manner to that described above for the von Mises stress plot. Users are encouraged to examine the displacement display and animate it on their own.

Analysis Insight

For those familiar with the actual hardware used in a trunion mounted hydraulic or pneumatic cylinder, the displacement animation should raise questions about the validity of loads applied to the pin. A photograph of a complete trunion mount, shown in Fig. 29, reveals that the central portion of the pin passes through a close fitting hole on the cylinder base similar to holes at both ends of the pin in the trunion base. A far superior model of the trunion mount and pin assembly should include additional geometry and a **Contact/Gaps** definition applied along the center portion of the pin.

Figure 29 – Close-up view of a trunion mount attached to the bottom of a pneumatic cylinder.

This insight is introduced to emphasize, once again, the significant influence of *proper boundary conditions* upon accurate finite element results. On your own, consider whether or not the accuracy of boundary conditions (i.e., **Loads** and **Restraints**) applied to this model could be further improved. For example, knowing that the trunion mount is to be bolted down, should another restraint be applied to the bottom surface of the trunion mount to prevent deflection *into* a rigid bottom surface? Also consider how additional restraints applied to the bottom surface of the trunion depend on the stiffness of the surface to which it is fastened.

Proper finite element analysis requires its users to carefully consider and model numerous factors that directly influence the validity of results.

Contact Pressure / Stress

The final section of this example explores use of the **Contact Pressure** plot. This plot displays, in a unique graphical format, the contact pressure developed between mating parts for which **Contact/Gaps** conditions are specified. Proceed as follows to display this plot.

1. Begin by turning off the display of loads and restraints on the model. Right-click the **Load/Restraint** folder and from the pull-down menu, select **Hide All**.

2. Next, right-click the **Results** folder and from the pull-down menu, select **Define Stress Plot…**. The **Stress Plot** property manager opens.

3. Within the **Display** dialogue box, click to open the stress **Component** pull-down menu and from the list of options, select **CP: Contact Pressure**.

4. Click **[OK]** ✓ to close the **Stress Plot** property manager. A new plot named **Stress2 (-Contact pressure-)** is listed beneath the **Results** folder.

5. If the contact pressure plot is not displayed, double-click **Stress2 (-Contact pressure-)**. *Examine the model carefully. The default display of contact pressure is relatively small.*

The following steps outline how to adjust the contact pressure plot to appear as illustrated in Figs. 30 (a) and (b).

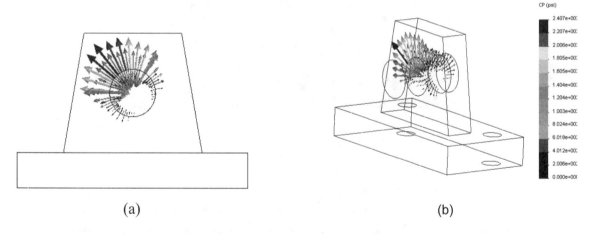

(a)   (b)

Figure 30 – (a) Left-side view of **Contact pressure** between the pin and hole in the trunion mount; (b) view showing three-dimensional nature of **Contact pressure** between mating surfaces.

6. Right-click **Stress2 (-Contact pressure-)** and from the pull-down menu select **Vector Plot Options…**. The **Vector plot options** property manager opens as shown in Fig. 31.

7. Within the **Size** field (upper spin-box in the **Options** dialogue box), type **600** to enlarge the contact pressure vectors displayed on the current plot. The value

chosen for the **Size** field is a user preference and should be selected to create a meaningful display.

8. Make certain that ⦿ **Match color chart** is selected. This option adds color to the vector plot such that, in addition to vector size indicating magnitude of contact pressure, the vector color also corresponds to that displayed in the accompanying color-coded legend.

9. Click **[OK]** ✓ to close the **Vector plot options** property manager. Your display should now appear similar to that shown in Figs. 30 (a) and (b).

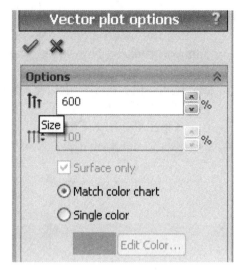

Figure 31 – Magnitude of the vector display is controlled within the **Vector plot options** property manager.

10. Rotate and zoom-in on the model to gain a better appreciation of the three-dimensional nature of contact pressure variation between the pin and hole in the trunion mount. Observe that higher contact pressure exists toward the loaded side (i.e., the middle) of the pin.

A **Contact Pressure** plot is somewhat unique because it displays both magnitude and direction of pressure between the contacting surfaces. Existence of this plot is dependent on the **Contact/Gaps** condition defined earlier in this example.

This concludes the analysis of contact pressure developed between the trunion base and a pin subject to an external load. This example file can either be saved or closed without saving at the discretion of the user. If file space is of a premium, delete the file.

11. From the main menu, select **File / Close** (or) **File / Save As** and proceed accordingly.

Analysis Insight

In closing, consider how methods of this chapter could have been applied to the pin used to attach the roller in the cam follower example of Chapter 1 or to determine contact stress in the vicinity of the pin hole in the curved beam model of Chapter 2.

Analysis of Machine Elements using COSMOSWorks

EXERCISES

1. At the push-rod end of a hydraulic or pneumatic cylinder (opposite end from the trunion mount) is a part that connects the push-rod to a driven component. This connector takes many forms, the most common, called a "clevis," is pictured in Fig. E 6-1. A clevis is typically threaded onto the end of a cylinder push-rod. It then is connected to a driven device by means of a pin. The figure below shows the geometry of a typical clevis with a reaction force, F = 8600 N, applied to its clevis pin. Open the files **Clevis 6-1** and **Clevis Pin 6-1** and perform a finite element analysis to determine the items requested below.

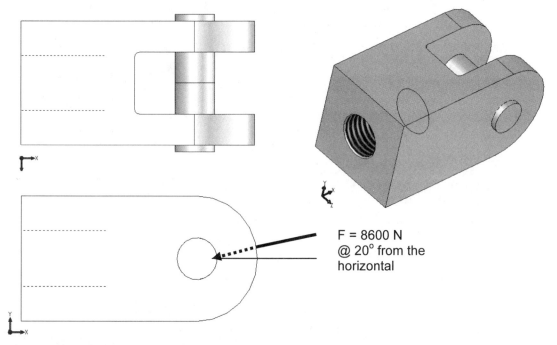

Figure E 6-1 – Top, front, and isometric views of a clevis and pin assembly. A force of 8600 N is applied to the pin as shown in the front view.

- Material: **Alloy Steel** (Use S.I. units)

- Mesh: **High Quality** tetrahedral elements

- Restraint: **Immovable** (appropriate for a push-rod threaded into the hole)
  Symmetry as necessary
  Contact/Gaps between pin surface and clevis hole, specify
  No penetration.

- Force: Apply "appropriate" force components to model the applied load

- Assumptions:
  ➢ The pin (40 mm long) is centered between sides of the clevis (clevis height and depth = 36 mm, square).
  ➢ Use symmetry to model half of the clevis and pin assembly.

Determine the following:

Develop a finite element model that includes: material specification, restraints, load(s), mesh, and a solution. Defeature the model as needed and delete the "cosmetic threads" as outlined below. *Reminder: If the example of the current chapter was worked, recall that system default units were changed to English at the outset of the example. Refer to section* "**Create a Finite Element Analysis (Study)**", *page 6-10 steps 5 to 7 to define SI units for the current exercise.*

Cosmetic Threads

So called "cosmetic threads" are included on the Clevis model to provide insight into its means of attachment to a cylinder push-rod end. Cosmetic threads are a symbolic representation of screw threads rather than actual geometric shapes cut into the model. As such, they serve their intended purpose of conveying information, but do not affect results. For example, they do not cause stress concentration due to thread profiles cut into the model. It is often desirable to **Delete** cosmetic threads because, if not deleted, they cause a circle and/or dashed lines to appear on various views of the model as shown in top, front, and isometric views in Fig. E 6-1. Although this remnant of the thread profile does not affect results, it is found bothersome by some individuals. To delete cosmetic threads, proceed as follows at the start of this exercise.

1. Open the file **Clevis 6-1**.

2. Toggle to **SolidWorks**, if not already in SolidWorks and, if necessary, double-click **Clevis 6-1** to display contents of the SolidWorks feature manager shown in Fig. E 6-2.

3. Near the bottom of the SolidWorks feature manger, click the "+" sign adjacent to **Threaded Hole Cut-Extrude2**.

4. Beneath this icon, right-click **Cosmetic Thread 4** and from the pull-down menu select **Delete**.

5. Return to the problem solution.

Figure E 6-2 – **SolidWorks** feature manager used to **Delete** the **Cosmetic Threads**.

a. Create a plot showing all restraints and loads applied to half of the assembly model. This image might be thought of as the finite element equivalent of a free-body diagram. Do not show stresses or a mesh on this plot. Adjacent to the image, hand write calculations used to determine the X and Y force components acting on the pin. Also, on a small sketch, oriented the same as the model, draw and label magnitudes of the X and Y force components applied to the *model*.

b. Create a plot showing **Contact Pressure** between the pin and clevis. Enlarge vector size to make vectors easily visible, but in reasonable proportion to the overall image. Select a view that clearly shows the three dimensional nature of contact pressure variation.

c. Create a plot of von Mises stress contours displayed on a deformed image of the pin and clevis assembly. Show the deformed model by checking ☑ **Deformed Shape** and select ⊙ **Automatic** as the scaling method. Also, incorporate automatic labeling of the maximum von Mises stress. See also part (d) below.

d. Create a von Mises stress plot showing regions of the assembly that have a safety factor less than 2.5.

e. Use **Iso Clipping** to create a plot that clearly shows regions of the model where von Mises stress exceeds the safety factor of 2.5. Include automatic labeling of the maximum von Mises stress on this plot.

f. Discuss **Restraints** applied to the clevis and pin model. Provide sound reasoning/ justification for restraints applied at the threaded clevis attachment location. If you believe the suggested restraints are incorrect or can be improved, provide a detailed discussion that justifies the opinion expressed.

g. Briefly describe the procedure used to plot the stress magnitudes for part (e) of this exercise and discuss the correspondence, or lack thereof, between the safety factor plot of part (d) and the Iso Clipping plot of part (e).

Textbook Problems
In addition to the above exercise, it is highly recommended that additional problems involving contact between mating parts be worked from a design of machine elements textbook. The parts need *not* be cylindrical. Textbook problems provide a great way to discover errors made in formulating a finite element analysis because they typically are well defined problems for which the solution is known. Textbook problems, if well defined in advance, make an excellent source of solutions for comparison.

CHAPTER #7

BOLTED JOINT ANALYSIS

Bolts, and/or nuts and bolts in combination, are one of the most frequently used means of joining mechanical components. Bolted joints are commonly used in non-permanent connections where access to and removal of components for repair or replacement is essential to the maintenance of mechanical devices. This example examines steps involved to successfully model bolted connections. Bolted connections are but one of many connection types available in COSMOSWorks. Other connector types not examined here include: rigid links, springs between component faces, elastic supports, pins, spot welds, and bearing between components.

Learning Objectives
Upon completion of this example, users should be able to:

- Define *bolt connectors*

- Define *custom material properties* for bolt connectors

- Identify when mesh refinement is necessary based on *high stress gradient*

- Define *contact sets* between mating parts without the use of Split Lines.

Problem Statement
An angle bracket is attached to a long, rigid support plate as illustrated in Fig. 1. Both the support plate and angle bracket are made of **ANSI 1020 Steel**. For purposes of this example, the support plate is shortened by arbitrarily cutting it in the vicinity of the two dashed lines on either side of the angle bracket. Four **M 12 x 1.75** bolts with nuts (i.e., 12 mm diameter metric bolts with a thread pitch = 1.75 mm) fasten the two parts together through the four holes shown. Bolts and nuts are not shown in the accompanying figure. All bolts are tightened to a preload of **24000** N. The goal is to determine bolt loads when the joint is loaded by a force of **5000** N acting normal to the top surface of the tab at the right end of the angle bracket. The force acts both toward and away from the tab surface shown on Fig. 1, but is not a cyclic load.

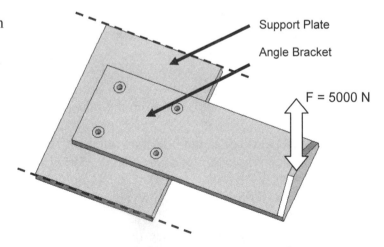

Figure 1 – Basic geometry of the angle bracket and support plate, showing bolt hole locations.

1. Open SolidWorks by making the following selections. (*Note:* "/" is used to separate successive menu selections.)

    Start / All Programs / SolidWorks 2008

2. When SolidWorks is open, select **File / Open**. Then use procedures common to your computer environment to open the COSMOSWorks file named "**Support Plate and Angle Bracket**."

Create a Static Analysis (Study)

1. Toggle to the **COSMOSWorks** manager by selecting its icon at top-right of the manager tree.

2. Right-click **Support Plate and Angle Bracket** and from the pull-down menu, select **Study…**. The **Study** property manager opens.

3. In the **Study name** field, type "**Bolted Joint-DOWNWARD Load**".

4. Accept the default **Solid Mesh** and **Static** analysis by clicking **[OK]** ✓ to close the **Study** property manager. An outline of the study is created in the COSMOSWorks manager.

Assign Material Properties to the Model

Material properties of the support plate and angle bracket are assigned in this section. Because both components are the same material, this procedure is very simple. Try it on your own. Steps are provided below if guidance is desired.

1. Right-click the **Solids** folder and from the pull-down menu, select **Apply Material to All…**. The **Material** window opens.

2. In the **Material** window, select ⊙ **From library files** and select **cosmos materials** in the pull-down menu. From the list of materials, click the "+" adjacent to **Steel (30)** and from the pull-down menu, choose **AISI 1020**.

3. In the right-half of the **Material** window, verify that **Units:** are set to **SI** and verify the material Yield Strength is **351571000** N/m^2.

4. Click **[OK]** to close the **Material** window.

Notice that bolt material is not specified at this time. Bolt material is specified independently when other bolt characteristics are defined.

Apply Load and Restraints

Traditional Loads and Restraints

This section outlines the application of restraints and an external load using procedures similar to those encountered throughout this user manual. Those wishing to apply immovable restraints to cut edges of the support plate and a downward 5000N load normal to the top-end of the angle bracket are encouraged to do so on their own. However, steps are provided below if guidance is desired.

1. Right-click the **Load/Restraint** folder and from the pull-down menu, select **Restraints…**. The **Restraint** property manager opens.

2. From the pull-down menu at top of the **Type** dialogue box, select **Immovable (No translation)** as shown in Fig. 2.

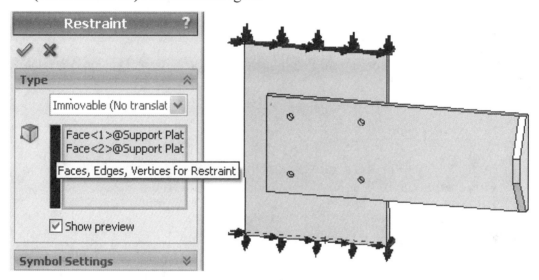

Figure 2 – Application of **Immovable** restraints to cut-surfaces on both ends of the support plate.

3. The **Faces, Edges, Vertices for Restraint** field is highlighted (light blue). Move the cursor into the graphics screen and rotate and zoom-in on the model as necessary to select the two cut surfaces of the support plate shown in Fig. 2. **Face<1>** and **Face<2>@Support Plate-1** appear in the **Type** dialogue box.

4. Click **[OK]** ✓ to close the **Restraint** property manager.

Next apply a downward force on top of the tab at the right-end of the angle bracket.

5. Right-click the **Load/Restraint** folder and from the pull-down menu, select **Force…**. The **Force** property manager opens as illustrated in Fig. 3.

6. At top of the **Type** dialogue box, click to select ⊙ **Apply normal force**.

7. The **Faces and Shell Edges for Normal Force** field is highlighted and awaits input. Proceed to select the top surface of the angle bracket shown in Fig. 3.

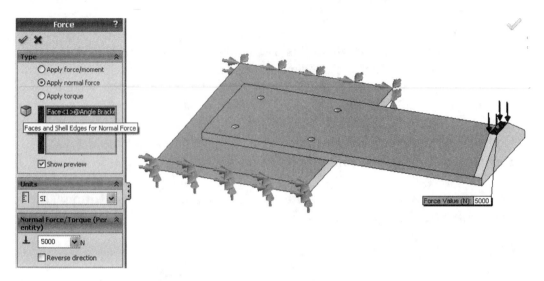

Figure 3 – Application of a downward force normal to the top surface of tab located at the right-end of the angle bracket.

8. In the **Units** dialogue box, verify that units are set to **SI**.

9. In the **Normal Force/Torque (Per entity)** dialogue box, type **5000** N in the **Force value** field. The force should act downward, if not check ☑ **Reverse Direction**.

10. Click **[OK]** ✓ to close the **Force** property manager.

Define Bolted Joint Restraints

Individual bolted fasteners used to join the angle bracket and support plate are defined next. Discussion below assumes the bolt-head is located on top of the angle bracket while the nut is located against the bottom surface of the support plate as identified in Fig. 4. Definition of bolted joints also takes place within the **Load/Restraint** folder.

Work through the following steps carefully and sequentially to define *one bolt at a time*. NOTE: If all bolt clamping surfaces are selected in a single step, software ability to isolate *individual* bolt reactions is lost. Also, while proceeding through the following steps, place the cursor onto each field or icon to reveal its name. This approach enhances insight into understanding the function of each option.

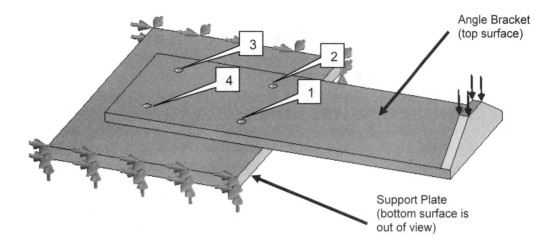

Figure 4 – Angle bracket and support plate showing bolt holes numbered from 1 to 4 in counterclockwise order.

1. Right-click **Load/Restraint** and from the pull-down menu select **Connectors…**. The **Connectors** property manager opens.

2. Begin by clicking the **Keep Visible** "push-pin" located at the top-right of the **Connectors** property manager circled in Fig. 5.

3. At top of the **Type** dialogue box, click to display the pull-down menu. From the list of possible connectors, select **Bolt**. Immediately, the **Connectors** property manager changes appearance to that shown in Fig. 6.

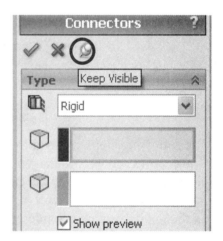

Figure 5 – Initial view of the **Connectors** property manager with the **Keep Visible** push-pin selected.

Within COSMOSWorks there exists two ways to define bolted connectors. The most direct method is outlined below. However, both methods presume the user has access to tables listing standard bolt proportions and material properties. In order to define a nut and bolt in COSMOSWorks, the parameters listed in Table 1, below, are needed. Values shown are applicable to grade 5.8 metric **M 12 x 1.75** bolts and nuts used in this example. Information in Table 1 is readily available in most design of machine elements texts.

Analysis of Machine Elements using COSMOSWorks

Table 1 – Values required to define nut and bolt characteristics.

PARAMETER	VALUE
Bolt shank diameter	12 mm
Bolt head diameter	18 mm
Nut diameter	18 mm
Bolt Elastic Modulus	207e9 N/m^2
Bolt Poisson's Ratio	0.292
Bolt Pre-load	24000 N

4. In the revised **Type** dialogue box, click to select **Standard or Counterbore with Nut** icon circled in Fig. 6. This action selects a standard nut and bolt for the connector.

Also move the cursor over the remaining icons to get a quick overview of other connector types within the **Bolt** connector sub-group.

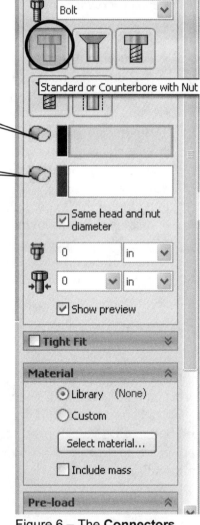

Circular edge of the bolt **HEAD** hole

Circular edge of the bolt **NUT** hole

The following steps proceed from top to bottom of the **Connectors** property manager. To ensure selection of the correct field, move the cursor over each field to reveal its name. Also recall that the bolt head is located on the angle bracket, considered to be the top of the model, and the nut is located on the support plate (bottom side) of the model in Fig. 8.

5. Click to activate (light blue) the **Circular Edge of The Bolt Head Hole** field (if not already selected). Then rotate and zoom-in on the *top* of the angle bracket and click to select the *top edge* of bolt hole #1; see Fig. 8. **Edge<1>@Angle Bracket-1** is listed in the active field of Fig. 8. A flyout table *may* appear adjacent to the selected bolt hole. This table summarizes all bolt parameters as they are defined. Click and drag the flyout table to a convenient location on the screen.

Figure 6 – The **Connectors** property manager after selecting **Bolt** as the connector type.

7-6

Bolted Joint Analysis

6. Next, in the **Type** dialogue box, click to activate (highlight) the **Circular Edge of The Bolt Nut Hole** field noted on Fig. 7. *CAUTION: This step is easily missed in the repetitive process outlined below.*

7. Rotate and zoom-in on the *bottom* of the support plate to select the corresponding bottom *edge* of bolt hole #1. **Edge<2>@ Support Plate-1** is listed in the active field.

8. Immediately beneath this field, check ☑ **Same head and nut diameter**. This step matches the nut and bolt head diameters as is common for standard bolt connections.

9. In the next field from the top, click to open the **Unit** pull-down menu and select **mm**. *Units must be specified prior to entering the bolt head diameter.*

10. To the left of the **Unit** field, type **18** mm in the **Head Diameter** field.

11. Next, adjacent to the **Bolt Shank Diameter** field, click to change **Unit** to **mm**.

12. In the **Bolt Shank Diameter** field, type **12** mm.

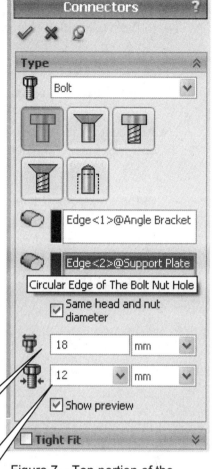

Figure 7 – Top portion of the **Connectors** property manager showing selections used to define bolts and nut contact faces.

13. Clear ☐ **Tight Fit** (if checked). This option does not pertain to bolts inserted through clearance holes such as are used in this example.

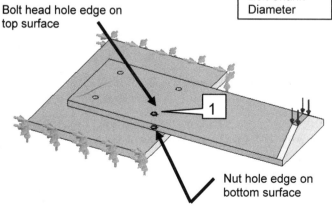

Figure 8 – Model showing selection of hole edges for the bolt head and nut contact faces for bolt #1.

Next, direct your attention to the **Material** dialogue box. In this dialogue box it is possible to select bolt material from the **Material** window, as has been done in all previous examples, or to specify custom material properties as outlined below.

14. In the **Material** dialogue box shown in Fig. 9, select ⊙ **Custom**. The bolt flyout table expands to display additional information to be specified by the user.

15. In the **Unit** dialogue box, select **SI** (if not already selected).

16. Adjacent to the **Young's Modulus** E_x field, type **207e9** N/m^2 to identify the bolt modulus of elasticity.

17. In the **Poisson's Ratio** field, type **0.292**.

18. Leave the **Thermal expansion coefficient** α blank.

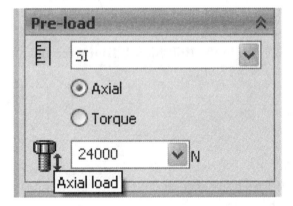

Figure 9 – Specifying custom material properties for bolts in the **Material** dialogue box.

Figure 10 – Bolt preload specified as an axial (tensile) force in the bolt.

19. Scroll to the bottom of the **Connectors** property manager where the **Pre-load** dialogue box appears as shown in Fig. 10. Open the **Unit** pull-down menu and from the list, again select **SI** (if not already selected).

20. Next, select ⊙ **Axial**. This selection indicates that the bolt preload is expressed by an axial (tensile) load in the bolt.

21. In the **Axial load** field, type **24000** N.

22. Return to the top of the **Connectors** property manager and click **[OK]** ✓. This action applies all of the above settings to bolt #1.

As specifications are entered for each bolt and nut, symbols representing bolt connectors appear at each hole location as illustrated in Fig. 11.

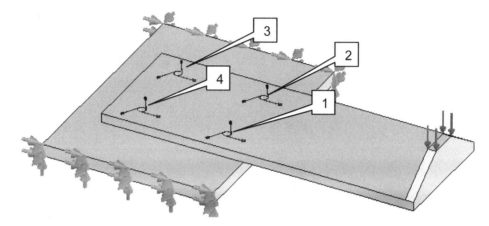

Figure 11 – Bolt connector symbols appear at locations of the Head and Nut Contact Faces. Rotate the model to view symbols on bottom of the support plate.

Because the **Keep visible** push-pin was selected, the **Connectors** property manager remains open and all values entered to describe the first bolt are retained. Unless changed by the user, the software assumes that additional identical bolts are used. Thus, it only remains to select the top and bottom hole-edges for the remaining bolts and nuts in the counterclockwise order shown in Fig. 11. Proceed as follows.

23. Within the **Type** dialogue box the **Circular Edge of the Bolt Head Hole** field should again be highlighted. On the *top* surface of the angle bracket, select the *edge* of hole #2. The hole edge is highlighted.

24. Beneath the above field, click to activate the **Circular Edge of the Bolt Nut Hole** filed. Rotate the model and zoom-in on the *corresponding* hole on the *bottom* of the support plate. Click to select the edge of this hole and bolt connector symbols appear on the model. Scroll down and notice that all other data entries remain as previously defined.

25. Return to the top of the **Connectors** property manager, and click **[OK]** ✓. This action applies the same settings to the currently selected bolt location.

Repeat steps 23 through 25 for *each* remaining bolt. Follow the counterclockwise order shown in Fig. 11.

26. After defining bolt #4 (and clicking **[OK]** ✓ in step 25), select **Cancel** ✗ to close the **Connectors** property manager.

At this point, **Bolt Connector-1** through **Bolt Connector-4** should be listed beneath the **Load/Restraint** folder in the COSMOSWorks manager tree as illustrated in Fig. 12.

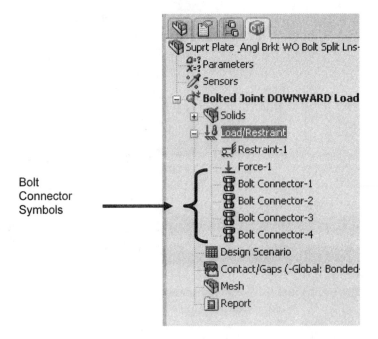

Bolt Connector Symbols

Figure 12 – Individually defined bolt connectors listed beneath the **Load/ Restraint** folder.

Analysis Insight

Instead of specifying bolt preload as an **Axial** force, in the **Pre-load** dialogue box, it is also possible to specify bolt preload in terms of torque applied to tighten the nut as shown in the **Pre-load** dialogue box of Fig. 13.

When bolt torque is specified, the software uses a rearranged form of equation [1], below, to compute axial preload. Notice that the torque coefficient **K**, listed as "**Friction Factor (K)**" at the bottom of the **Pre-load** dialogue box, can be changed to account for specific joint conditions.

A variety of joint and bolt characteristics, such as: surface finish (painted, plated, clean, etc.); alignment (parallelism of surfaces and/or flatness under the bolt head and nut); and thread condition (dry, lubricated, coated, etc.) all affect the torque coefficient **K**.

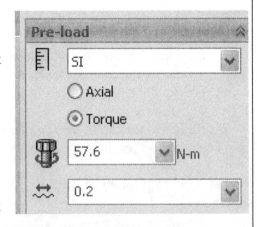

Figure 13 – **Preload** dialogue box showing specification of bolt **Torque** rather than an axial preload value.

> The relationship between bolt preload and bolt torque is typically defined in a design of machine elements text by the following equation.
>
> $$T = KF_id \qquad [1]$$
>
> Where: F_i = desired bolt preload (the initial axial tensile force in the bolt)
> T = applied torque required to develop a desired preload F_i
> K = torque coefficient (K = 0.20 is a generally accepted value)
> d = nominal bolt diameter
>
> For this example, if ◉ **Torque** were selected rather than ○ **Axial** preload, then the torque used to tighten the bolt is given by:
>
> $$T = KF_id = (0.20)(24 \text{ kN})(0.012 \text{ m}) = 57.6 \text{ N-m}$$
>
> See the entry typed into the **Torque** field in Fig. 13.

This completes the definition of individual bolt connectors. However, because we are dealing with an assembly, it is also necessary to define contact conditions between the two members that are bolted together. This task is addressed in the following section.

Define Local Contact Conditions

Contact between the angle bracket and support plate must also be defined. Because there are only two components in this assembly, it would be possible to define contact between mating surfaces on these two parts using a *Global* contact specification (i.e., a specification that applies to all contacting surfaces). However, it is more instructive to define contact in a *Local* sense (i.e., specific to the two contacting surfaces). This approach provides the user with greater control when defining contact situations and prepares the user to deal with more unique situations should they be encountered in other modeling applications. Begin by switching to an exploded view.

1. In the main menu, select **Insert** and from the pull-down menu choose **Exploded View…**. The Explode property manager opens.

2. Click to select the angle bracket (top part) in Fig. 14. A coordinate system triad appears on the part. Drag the Y-axis of the triad upward. Release the mouse button when the figure looks similar to Fig. 14. (Alternatively, drag the support plate downward).

Figure 14 – Exploded view prior to defining contact conditions.

3. Click **[OK]** ✓ to close the **Explode** property manager.

4. Right-click the **Contact/Gaps (-Global: Bonded-)** folder and from the pull-down menu select **Define Contact Set…**. The **Contact Set** property manager opens as shown in the middle of Fig. 15.

5. In the pull-down menu at top of the **Type** dialogue box, select **No Penetration**.

6. The **Faces, Edges, Vertices for Source** field is highlighted (light blue) and awaits input. Move the cursor into the graphics screen and rotate the model to select the bottom, left-end of the angle bracket shown highlighted in Fig. 15 (a). **Face<1>@Angle Bracket-1** is listed in the active field. A *Split Line* is provided to facilitate selecting only the contact surface.

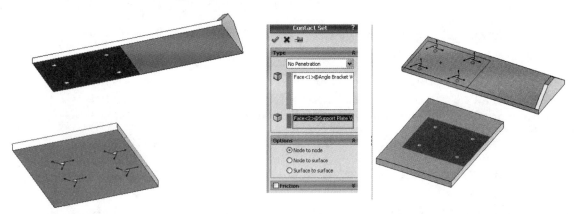

(a) Select bottom of Angle Bracket.   (b) Select top of Support Plate.

Figure 15 – Selection of equal and opposite faces defined to be in contact with **No Penetration**.

7. Click to highlight (light blue) the **Faces for Target** field, located at the bottom of the **Type** dialogue box. Then, proceed to select the contact face in the vicinity of bolt holes on top of the support plate. See highlighted face in Fig. 15 (b). **Face<2>@ Support Plate-1** is listed in the active field. Once again *Split Lines* are provided to permit selection of only the contact surface.

8. In the **Options** dialogue box, click to choose ⊙ **Node to surface**.

Analysis Insight:

a. It is also possible to select either **Node to node** (the least accurate) or **Surface to surface** because all of these contact options are valid when **No penetration** is selected. Each option allows surfaces to separate, which is important later in this analysis. The **Node to surface** option is faster than the other two options. It also has the ability to model line or point contact while the **Surface to surface** contact option is more accurate, but requires more computation time and resources. The user is encouraged to use COSMOSWorks **Help** for additional information.

> b. Before closing the **Contact Set** property manager, click to place a check mark in the ☑ **Friction:** dialogue box. This dialogue box reveals that it is possible to define friction between the contacting surfaces. This option is typically applied when the potential for movement is investigated between mating parts. Since that is not the case in this example, clear the ☐ **Friction** check-box before proceeding.

9. Click **[OK]** ✓ to close the **Contact Set** property manager. Highlighting on the model disappears and **Contact Set-1 (-No Penetration<Angle Bracket-1, Support Plate-1>-)** is listed beneath the **Contact/Gaps** folder in the COSMOSWorks manager tree.

Before proceeding, the model is returned to its original, un-exploded, state as follows.

10. Toggle to the **Configuration Manager** by clicking its icon at top of the COSMOSWorks manager tree.

11. Click the "+" sign adjacent to **Default<Default Display State-1> [Support Plate and Angle Bracket]**.

12. Right-click ExplView1 and from the pull-down menu, select **Collapse**.

13. Toggle back to the **COSMOSWorks** analysis manager.

Mesh the Model and Run Solution

1. Right-click the **Mesh** folder and from the pull-down menu select **Create Mesh…**. The **Mesh** property manager opens.

2. Within the **Mesh Parameters** dialogue box, set **Unit** to **mm**, and accept the default mesh size.

3. Click to open the **Options** dialogue box and verify the following settings.

    - **Quality:** ⊙ **High**

    - **Mesher:** ⊙ **Standard**

    - **Mesher options:** ☑ **Jacobian Check for solid [4 Points]**

7-13

4. Return to the **Mesh Parameters** dialogue box and check ☑ **Run analysis after meshing**. Then, click **[OK]** ✓ to close the **Mesh** property manager.

Notice the significant increase in time required to solve this problem (approximately five minutes on a Pentium 4 PC). The primary increase occurs during the **Solving contact constraints:** portion of the solution where contact conditions between mating parts are determined.

Results Analysis (Downward Load)

Von Mises Stress

Briefly examine a plot of von Mises stress to gain an understanding of the magnitude and distribution of this stress throughout the assembly.

1. Click the "+" sign adjacent to the **Results** folder (if not already selected).

2. Right-click **Stress1 (-vonMises-)** and from the pull-down menu, select **Show**. A plot of von Mises stress is displayed on the model.

3. Right-click **Stress1 (-vonMises-)** and from the pull-down menu select **Edit Definition...**. The **Stress Plot** property manager opens.

4. In the **Display** dialogue box change **Units** to **N/m^2**.

5. In the **Advance Options** dialogue box, clear the check mark from ☐ **Average results across boundary for parts**.

6. At top of the **Deformed Shape** dialogue box, verify that ☑ **Deformed Shape** is checked. Also within this dialogue box, select ⊙ **Automatic** to apply the system default exaggerated deformation to the model display.

7. Click **[OK]** ✓ to close the **Stress Plot** property manager.

8. Right-click **Stress1 (-vonMises-)** and from the pull-down menu select **Chart Options...**. The **Chart Options** property manager opens.

9. In the **Display Options** dialogue box, check both ☑ **Show min annotation** and ☑ **Show max annotation**.

10. Click **[OK]** ✓ to close the **Chart Options** property manager.

A plot of von Mises stress on the deformed model should appear as illustrated in Fig. 16.

Observe that the maximum von Mises stress, indicated on the model and in the color-coded legend as 2.095e+008 N/m^2, is considerably less than the material Yield Strength of 3.5157e+008 N/m^2.

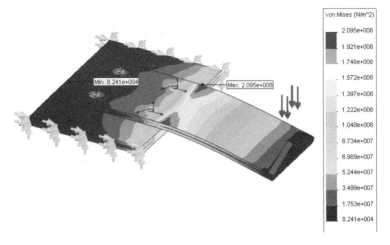

Figure 16 – von Mises stress distribution as seen in a top view of the assembly corresponding to a downward acting load on the model.

Bolt Forces

A primary feature of a bolt analysis is the ability to determine forces acting at each bolt. In the assembly analyzed in this example, it is reasonable to expect similar bolt forces in bolts #1 and #2, which are equidistant from the applied load. Similarly for bolts #3 and #4, which are located a further, but equal, distance from the applied load. Axial bolt force is the algebraic sum of its preload plus effects due to external load(s). Bolt forces are examined as follows.

1. Right-click the **Results** folder, and from the pull-down menu, select **List Pin/Bolt /Bearing Force...**. The **Pin/Bolt/Bearing Force** window opens as shown in Fig. 17.

Examination of this window reveals the **Study name:** in the upper left corner. Beneath the study name is a pull-down menu where it is possible to select results for **All Pins**, **All bolts** or results for each individual bolt. From this pull-down menu, select **Bolt Connector-1**. Adjacent to the bolt number is the current set of units. Set **Units:** to **SI**, if necessary, for the current example. Finally, at the bottom of the window, the **X**, **Y**, **Z**-components and the **Resultant** are listed for the **Shear Force (N)**, **Axial Force (N)**, and **Bending moment (N-m)** that act on the bolt.

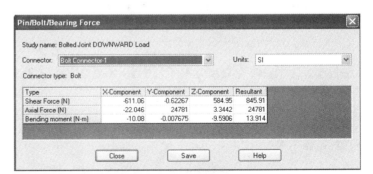

Figure 17 – The **Pin/Bolt/Bearing Force** window summarizes all Shear Force, Axial Force, and Bending moments that act on each bolt in the joint.

2. In the **Axial Force** row, observe the **Resultant** in the far right column. Its value is **24781** N. *Note: Your values may differ slightly due to computational round-off.*

3. Click the **Connector:** pull-down menu and from the list of connectors, select **Bolt Connector-2**. This bolt is positioned adjacent to bolt #1 on the model. Observe that the **Resultant** bolt force is **24793** N.

4. Next, display results for **Bolt Connector-3** and **Bolt Connector-4**. Again observe values of the **Resultant** axial bolt forces. For bolt #3, Resultant = 23951 N, and for bolt #4, Resultant = 23958 N.

5. Close the **Pin/Bolt/Bearing Force** window by clicking the **[Close]** button.

Observations

a. Forces at both bolts in the same row are nearly identical in magnitude (allowing for some computational inaccuracy)

b. Forces at bolts #1 and #2 differ at most by 3.3 % from the specified bolt preload (24000 N) and both loads are slightly *higher* than the given preload.

c. Note the small difference (only 7 N) between forces in bolts #3 and #4 that are located in the row furthest from the applied load. These values differ by 0.18 % from the specified bolt preload of 24000 N. But, due to joint loading, these values are both slightly *less* than the original bolt preload.

d. The bolted joint is well designed because most of the load is carried by the material clamped between the bolt head and nut. This observation is based on the fact that bolt preload differs only slightly from the design value of 24000 N.

e. Symmetry of results between bolts in the two different rows lends credence to the fact that the problem is formulated correctly. That is to say, loads on bolts #1 and #2 are nearly the same while the same is true for loads on bolts #3 and #4. However, loads on bolts at different distances from the applied load are observed to differ from one another.

Define a New Study with the Applied Load Reversed

Because the above analysis does not reveal significantly different results at the two rows of bolts, the example is next re-worked with an upward load applied to the angle bracket as shown in Fig. 18. Proceed as follows.

1. Make a copy of the current study by right-clicking the **Bolted Joint-DOWNWARD Load (-Default-)** study. From the pull-down menu, select **Copy**.

2. At top of the COSMOSWorks manager tree, right-click **Support Plate and Angle Bracket**. From the pull-down menu, select **Paste**.

3. The **Define Study Name** window opens. In the **Study Name:** field, type **Bolted Joint-UPWARD Force** and click **[OK]**. This action creates an exact copy of the former study. If necessary, click the "+" sign adjacent to **Bolted Joint-UPWARD Force** in the COSMOSWorks manager.

To differentiate this study from its predecessor, it is only necessary to reverse the load applied to the angle bracket. To accomplish this, proceed as follows.

4. If necessary, click the "+" sign adjacent to **Load/Restraint** under the UPWARD study and right-click the **Force-1**. From the pull-down menu select **Edit Definition…**. The **Force** property manager opens.

5. At the bottom of the **Normal Force/Torque (Per entity)** dialogue box, click to check ☑ **Reverse direction**.

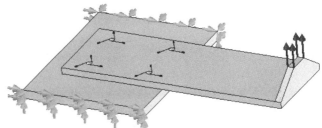

Figure 18 – Bolted joint model with the applied force reversed so that it acts upward.

6. Click **[OK]** ✓ to close the **Force** property manager. The load applied to the right-end of the model should now be directed upward as shown in Fig. 18.

Because all aspects of the previous study were copied, nothing else need be done except to re-run the solution due to the altered loading condition. Re-running the solution eliminates the *warning* ⚠ symbol adjacent to the **Results** folder and the *error* ⬇ symbol adjacent to the **Bolted Joint-UPWARD Force** study name. Re-run the solution on your own, or execute the following step.

7. Right-click the **Bolted Joint-UPWARD Force (-Default-)** folder and from the pull-down menu, select **Run**.

Results for the bolted assembly subject to an upward acting load are examined below.

Results Analysis (Upward Load)

Von Mises Stress
Begin by examining von Mises stress.

1. Click the "+" sign adjacent to the **Results** folder.

2. Double-click **Stress1 (-von Mises-)**. The plot illustrated in Fig. 19 appears. Note the small circled areas of high stress.

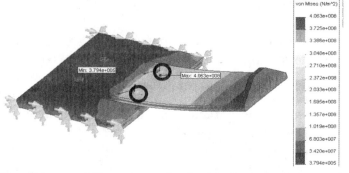

Figure 19 – von Mises stress distribution throughout the model corresponding to an *upward* acting load of 5000 N.

To obtain a good overview of stress distribution throughout the model, rotate the model to view all surfaces. Notice that, corresponding to an upward load, maximum stress within the model 4.063e008 N/m^2 (*values were observed to vary between 4.063e+008 N/m^2 to 3.992e+008 N/m^2 in subsequent analysis runs*) exceeds the material Yield Strength of 3.5157e+008 N/m^2 by approximately 15.5%. Close examination reveals the region of maximum stress is located adjacent to bolt hole #2 at the interface between the two parts. The region of maximum stress is exceedingly small and some ramifications of this finding are discussed below. The interested reader is encouraged to zoom in on the area of maximum stress and to apply iso-clipping, as outlined in Chapter 6, to view the high stress gradient in this region.

Analysis Insight

If the load is statically applied and the material is ductile, it can be argued that localized yielding occurs in the maximum stress region. A further argument is that localized yielding in the highly stressed area results in increased strength. This increase is due to the phenomenon commonly referred to as "cold-working" of the material. On the other hand, if the material is brittle or if loads are repeatedly applied and released (or reversed) failure by fracture or fatigue will likely occur.

The most instructive way to investigate any high stress region is with the mesh displayed on the model. The reason for this is that the rate of change of stress magnitude, also known as "stress gradient," can be observed in relation to element size. However, the three dimensional nature of the mesh makes it difficult to obtain easily interpreted textbook images in the region of maximum stress. Therefore, we next examine a region of high stress gradient on the surface of the angle bracket to demonstrate its significance relative to mesh size.

Begin by observing the two small *areas* of high stress that occur adjacent to bolt #1 and bolt #2 circled in Fig. 19.

To further emphasize the stress gradient in this region, Fig. 20 shows an enlarged view of the area adjacent to either bolt #1 or #2 with a mesh superimposed on the model. On your own, create a view like that shown in Fig. 20. The following steps are provided in the event that guidance is desired.

3. Right-click **Stress1 (-von Mises-)**. From the pull-down menu, select **Settings…** The **Settings** property manager opens.

4. Verify that the **Fringe Options** dialogue box is set to **Discrete**; if not change it. This action further emphasizes the small boundaries of the high stress (yellow) area relative to mesh size.

5. From the pull-down menu in the **Boundary Options** dialogue box, select **Mesh**. A mesh is superimposed on the model.

6. Click **[OK]** ✓ to close the **Settings** property manager.

7. Zoom-in on either of the high stress areas circled on Fig. 19 to obtain an image similar to Fig. 20. The high stress area is outlined for emphasis.

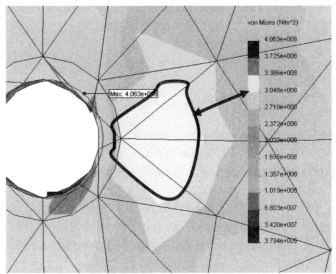

Figure 20 – Enlarged view showing the maximum stress region lies within a *portion* of the elements circled in Fig. 20. Stress magnitude is correlated to the color-coded stress scale.

Conclusion: Because the high stress region is *smaller* than any individual element (i.e., it occupies only a portion of each element), it is desirable to use a smaller mesh in this region. Recall that a finer mesh reveals stress levels more accurately. Because this small area occurs adjacent to a geometric discontinuity (the bolt hole), use of *Mesh Control* outlined in Chapter 3, should be considered.

The above observation is of paramount importance in any analysis where a high stress gradient occurs. However, because determination of bolt forces rather than maximum stress in the model is the primary goal of this example, we return to the analysis of bolted joint results without applying mesh refinement at this time.

Bolt Forces (Upward Load)

The effect of altering direction of the applied load upon bolt forces is examined next.

1. Right-click the **Results** folder, and from the pull-down menu, select **List Pin/Bolt /Bearing Force…**. The **Pin/Bolt/Bearing Force** window opens.

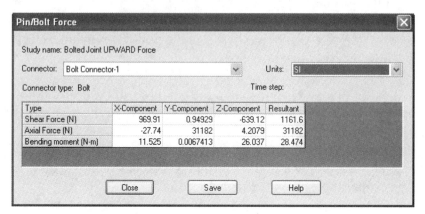

Figure 21 – Forces at **bolt #1** corresponding to an upward load applied to the angle bracket.

2. Verify that **Units:** are set to **SI**. If not, change to **SI** units.

3. Open the pull-down menu adjacent to **Connector:** and from the list of connectors, select **Bolt Connector-1**. The **Pin/Bolt/Bearing Force** window should appear as shown in Fig. 21. (Alternatively, select **All Bolts**, and scroll through the results.)

4. Observe the **Axial Force (N)** in the **Resultant** column at the far right. Its value is **31182** N (your value may differ slightly). This value represents an increased axial (tensile) bolt load of 6396 N above that for the case of the downward load applied to the angle bracket

5. Open the pull-down menu adjacent to **Connector:** and from the list of connectors, select **Bolt Connector-2**. In the **Axial Force** row, the **Resultant** force in this bolt is now **31110** N, which is reasonably close (within 72 N) to the value obtained at bolt #1 in the previous step.

Recall that bolts #1 and #2 are located adjacent to one another at the same distance from the applied load. Theoretically, identical bolt loads are expected at these two locations. However, due to inaccuracies and round-off error, the small differences noted above are considered acceptable.

6. Repeat steps 3 and 4 for **Bolt Connector-3** and **Bolt Connector-4**. Axial bolt forces at these two locations are **24047** N and **24039** N respectively.

A side view of the model in Fig. 22 supports the finding of larger bolt axial forces at bolts nearest to the applied load due to the additional upward "pull" resisted by bolts #1 and #2. Likewise, only a slight increase is expected at bolts #3 and #4, which are farthest from the applied load.

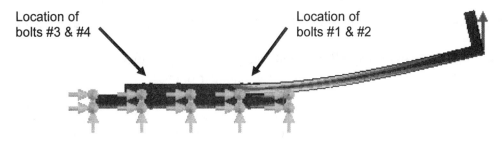

Figure 22 – Side view of the bolted joint with the angle bracket subject to an upward load at its right end.

Analysis Insight

Before closing the **Pin/Bolt Force** window, observe magnitudes of the **Shear Force** and the **Bending moment**. Whereas, **Bending moment** in the bolt is so small as to be almost negligible, the **Shear force** magnitude varies from a high of 1161.6 N at bolts #1 and #2 to a low of 917.14 at bolts #3 and #4 respectively. Shear forces, while not negligible, are small compared to axial forces in the bolts. More importantly, however, shear stress in the bolt subject to the maximum shear force is small relative to bolt shear strength. The following calculation examines shear stress in the most heavily loaded bolt.

$$\tau = F_s/A_s = (1161.6 \text{ N}) / (0.0763 \text{ m}^2) = 15.22 \text{ k N/m}^2 = 0.01522 \text{ MPa} \qquad [2]$$

where:
F_s = maximum shear force at a bolt
A_s = minor area of screw thread (at root of thread profile) NOTE: For structural bolts, the nominal bolt diameter should be used.
τ = shear stress in bolt caused by maximum shear force

When compared to a conservative value of shear strength (S_{sy} = 234 MPa) for a Grade 5.8 metric bolt, a safety factor of approximately 15 is found. For this reason, shear stress is not considered significant in this example.

However, for many structural applications, shear in bolts is the *primary* mode of loading. In those instances, shear load on bolts is of prime importance. An end-of-chapter exercise examines a bolted joint where shear in the joint is the governing factor.

Analysis of Machine Elements using COSMOSWorks

It may be possible to gain some insight into the origins of the X and Z components of shear force, by animating the model. To do this, proceed as follows.

7. Select **[Close]** to exit the **Pin/Bolt/Bearing Force** window.

8. If the von Mises stress plot is not displayed, double-click **Stress1 (-vonMises-)**.

9. Right-click **Stress1 (-vonMises-)** and from the pull-down menu, select **Edit Definition…**. The **Stress Plot** property manager opens.

10. In the **Deformed Shape** dialogue box, make sure ☑ **Deformed Shape** is checked. In the same dialogue box, notice that **Automatic** is selected and the distortion scale factor is **7.9573** (values may differ slightly). This value is altered below.

11. Click **[OK]** ✓ to close the **Stress Plot** property manager.

12. Right-click **Stress1 (-vonMises-)** and from the pull-down menu, select **Animate…**. If the model is animated, select the **Stop** ■ icon.

13. In the **Basics** dialogue box, increase the **Frames** to **10** and move the **Speed** control slide to the left. These actions smooth-out and slow the animation.

14. Click the **Start.** ▶ icon.

15. Slowly rotate the model to examine its deformation due to the applied load. Pay particular attention to deformations in the region near bolts #1 and #2 where the angle bracket and support plate intersect.

16. Click **[OK]** ✓ to close the **Animation** property manager.

Because it may be difficult to observe deformations that cause shear forces in bolt #1 and bolt #2, steps 9 through 15 above are repeated, except replace step 10 with the following.

17. In the ☑ **Deformed Shape** dialogue box, select ⦿ **User Defined**. In the adjacent field, type **30**. Then continue at step 11. This action increases the magnitude of the deformation displayed.

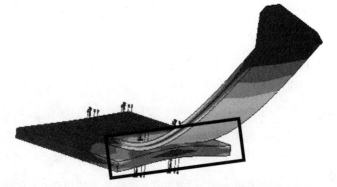

Figure 23 – Using animation to visualize origins of shear forces in the X and Z-directions at bolts #1 and #2.

The deformed model should appear similar to that illustrated in Fig. 23. In the following discussion, particular attention is focused on deformations between mating parts in the boxed region on this figure.

18. **Stop** ■ the animation and click **[OK]** ✓ to close the **Animation** property manager.

Figure 24 (below) represents a top view of the bolted joint where bolt connectors are represented by circles. Shear force components, directed according to their ± signs listed in the **Pin/Bolt/Bearing Force** connectors window, are shown at each bolt location. Using a combination of Fig. 24 and deformations viewed when animating the model, directions of the shear force components should begin to "make sense."

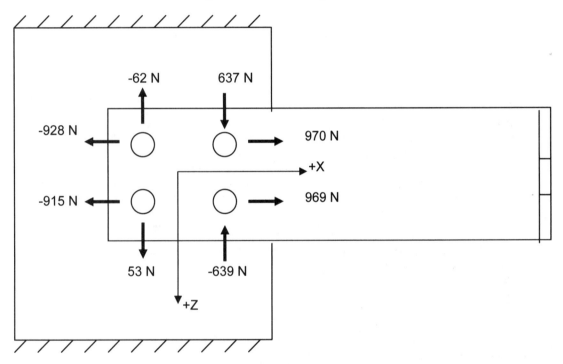

Figure 24 – Shear force components in the X and Z-directions developed in each bolt.

Bolt Clamping Pressure

Bolt preload is known to apply an equal and opposite clamping force on the joined members. Using software capabilities investigated in the previous chapter, contact pressure is investigated next. This analysis provides further insight into the intent of the bolted joint analysis capability within COSMOSWorks.

1. Right-click the **Results** folder and from the pull-down menu, select **Define Stress Plot...** The **Stress Plot** property manager opens.

2. In the **Display** dialogue box, click to open the pull-down menu adjacent to **VON: von Mises Stress** and from the pull-down menu select **CP: Contact Pressure**.

3. Verify that **Units** are set to **N/m^2**.

4. Next, in the **Advanced Options** dialogue box, clear the check-mark from ☐ **Average results across boundary for parts**.

5. Click **[OK]** ✓ to close the **Stress Plot** property manager. A new plot, labeled **Stress2 (-Contact pressure-)** is listed below the **Results** folder.

6. If not already displayed on the graphics screen, double-click **Stress2 (-Contact pressure-)** to display a contact pressure plot similar to that shown in Fig. 25.

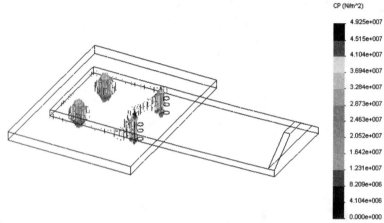

Figure 25 – Contact pressure plot showing interaction between the angle bracket and the support plate.

The following steps outline procedures to alter appearance of the graphic display in the event it does not initially look like Fig. 25. Users are encouraged to adjust the image on their own. However, steps are provided below if guidance is desired.

7. Right-click **Stress2 (-Contact pressure-)** and from the pull-down menu, select **Settings…**. The **Settings** property manager opens.

8. Within the **Boundary options** dialogue box, click to display the pull-down menu and from it select **Model**. An outline of the model appears. The **Fringe options** dialogue box can be set to either **Discrete** (preferred) or **Continuous**.

9. Click **[OK]** ✓ to close the **Settings** property manager.

Because restraints or bolt restraint symbols might obscure other pertinent graphical information, temporarily remove them from the display as follows.

10. In the COSMOSWorks manger tree, right-click **Load/Restraint** and from the pull-down menu select **Hide All**.

If the contact pressure vectors initially appear too small, which is often the case, increase their size as follows. Try this on your own or follow the steps listed below.

11. Double-click **Stress2(-Contact pressure-)** to again display the contact pressure plot.

12. Right-click **Stress2 (-Contact pressure-)** and from the pull-down menu select **Vector Plot Options…**. The **Vector plot options** property manager opens.

12. In the **Options** dialogue box, type **800** in the **Size** field (upper spin-box) to enlarge the size of contact pressure vectors displayed. Vector **Size** is a user preference.

13. Accept other default settings in this property manager and click **[OK]** ✓ to close it. Your display should now appear similar to Figs. 25 and 26.

14. Rotate and zoom-in on the model to gain better insight into what is actually plotted.

Figure 26 shows three different views of contact pressure vectors in the vicinity of the bolt holes and throughout the assembly.

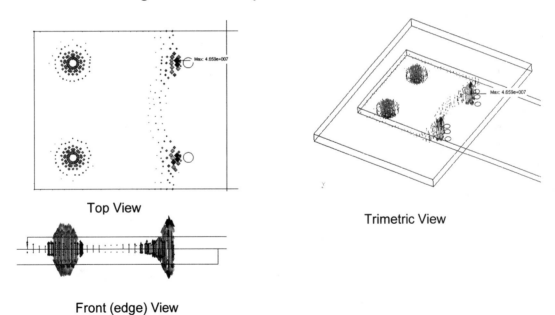

Figure 26 – Various views showing distribution of contact pressure between mating parts in the bolted joint subject to an upward load.

Analysis Insight

Do the graphical results in Fig. 26 (currently displayed on your screen) correspond with what you expected to see? The answer might surprise many users. However, it is valuable to understand what these plots reveal in order to ensure that results are interpreted correctly. Some additional questions that might be asked are:

a. Is the contact pressure shown in Fig. 26 applied by the bolts onto the plates?

or

b. Is the contact pressure shown in Fig. 26 the pressure that exists between the plates themselves?

or

c. Do the plots represent the superposition of contact pressure due to bolt clamping forces (i.e., forces beneath the head of the bolt and nut) and due to plate contact?

The answer to question (a) is that, although bolt preload "grips" the parts together, a **Contact/Gaps** condition was not specified between contacting faces of the bolt or nut and corresponding surfaces on the parts. For this reason, contact pressure between the bolt or nut and the plate surface is *not* illustrated.

Due to the above observation, it should be clear that the answer to question (b) is that contact pressure between the mating parts (i.e., between the plates) *is* shown in Fig. 26. Verify this by observing the direction of pressure vectors shown in the front view of the contacting surfaces. All vectors originate at the surface of each part and point away from the surface on which pressure acts. Also, in all views, notice that no contact pressure is shown to the right of bolts #1 and #2. This is consistent with Fig. 22, repeated below, which indicates a location where potential separation (i.e., a gap) between the plates may develop. This gap would reduce contact pressure between the structural members.

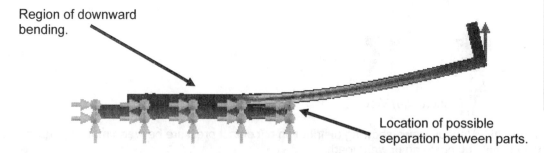

Figure 22 – (repeated) Probable location for a **Gap** to develop between parts.

Figure 26 also reveals a "footprint" of pressure, in the form of an arc, created by downward bending of the bracket to the left of and between bolts #1 and #2 caused by the upward deformation of the part to the right of these bolts.

Analysis Insight (continued)

Question (c) is already answered in response (a) above. However, it is worthy to note that even if no external load were applied to the angle bracket, contact pressure would still result in both plates around all four bolt connectors. In other words, the images show that bolt preload causes contact pressure between the parts that are gripped rather than between the bolt connectors themselves and the gripped members.

Bolts #3 and #4 are relatively unaffected by the upward load applied at the right end of the bracket. Therefore, notice the fairly uniform shape of contact pressure distribution both in terms of magnitude (front view) and geometry ("ring" shapes in top view) around these bolt holes.

Summary

Based on this example, it is evident that bolt force determination is one of the primary outcomes of a bolted joint analysis. This capability is useful for predicting and verifying design or analysis of bolt connections under load. Further, application of the **Contact Pressure** capability permits an analyst to determine contact pressure distribution in mating parts in the vicinity of connectors, such as bolts, thereby enhancing understanding of factors such as bolt spacing. Also, exercises 1 and 2 below outline a different method for defining contact between mating parts that does not require the use of Split Lines. This concludes the study of bolted joints, their definition, resulting contact pressure(s), and bolt reaction forces.

This example file can either be saved or closed, without saving, at the discretion of the user.

1. From the main menu, select **File / Close** (or) **File / Save As** and proceed accordingly.

EXERCISES

1. A steel plate is connected to a vertical column by means of four **M 16 x 2, Grade 8.8** steel bolts. Bolt material properties are: **E = 207 GPa** and **Poisson's Ratio = 0.292**. The column is a 254 mm x 76 mm structural steel channel. Both the column and plate are made of **ASTM A36 Steel**. The channel can be considered rigidly fixed (**immovable**) above and below the connection location; only a segment of the column is shown in the figures below. The plate is subject to a vertical load of **16 kN** applied to its top-right *edge* as shown in Fig. E 7-1. Figure E 7-1 shows the basic geometry of parts in this assembly. For simplicity, detailed images of the bolts are not shown. Open files **Column 7-1** and **Plate 7-1** and perform a finite element analysis to determine items requested below.

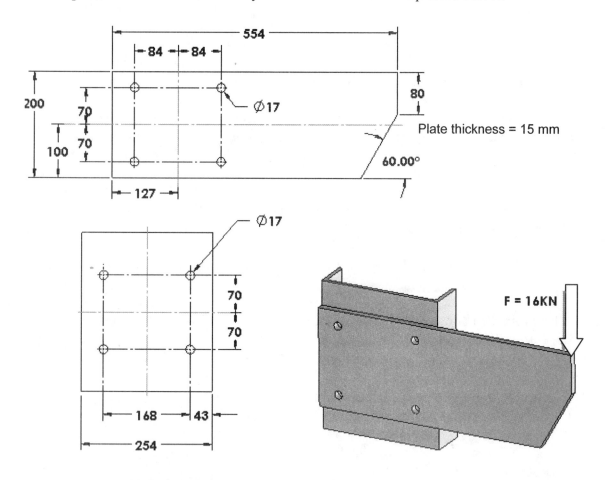

Column Segment (web thickness = 8 mm)

Figure E 7-1 – Basic geometry of the column and plate. A 16 kN load is applied on upper-right *edge* of the plate.

Bolted Joint Analysis

- Material: **ASTM A36 Steel** (column and plate material properties)
  E = 207e9 Pa, and Poisson's ratio **ν = 0.292** (bolt properties)

- Mesh: **High Quality** tetrahedral elements

- Restraints: **Immovable** as necessary on top and bottom cut surfaces of the column.

  Contact/Gaps with no penetration between the column and plate. A simplified method of defining contact between these two parts is outlined below.

  Bolt connector with **Nut**, use a washer diameter = 30 mm under the bolt-head and nut.

- Force: **16 kN** downward at top-right *edge* of the plate.

- Assumptions: Bolts fit in clearance holes where D_{Hole} = 17 mm
  Bolts are field tightened to a torque T = 226 N-m
  Bolt torque coefficient K = 0.20

Determine the Following:
a. Begin by creating an assembly of the **Column 7-1** and **Plate 7-1** parts; see **Solution Guidance** below. Then develop a finite element model that includes: material specification, restraints, load(s), bolt connectors, mesh, and solution.

Solution Guidance

This section outlines a simplified means for defining contact between mating parts *without* the need for *Split Lines*. Perform steps below after creating an assembly of the **Column** and **Plate** and after defining **Mates** between these two parts. The model should appear as shown in Fig. E7-2.

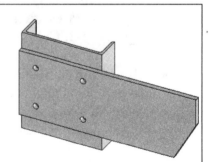

Figure E7-2 – **Column** and **Plate** assembly prior to definition of contact conditions.

The following steps outline use of the **Find Contact Sets** capability to assist in the definition of contact characteristics between mating parts. Despite the power of this option, it should be used with care because, in complex assemblies, it may find extra contact sets that are not intended or it may not find a contact set that the user wants. As a case in point, the **Find Contact Sets** option is applied *prior* to defining bolted connectors. This is done because bolted connectors add the potential for contact under the bolt head and nut and between the bolt shank and the inside of the bolt holes. This procedure eliminates the possibility of unwanted contact sets being defined.

Analysis of Machine Elements using COSMOSWorks

Solution Guidance (continued)
Proceed as follows to apply the simplified contact set capability.

- In the COSMOSWorks manager, right-click the **Contact/Gaps (-Global: Bonded-)** icon and from the pull-down menu, select **Find Contact Sets....** The **Find Contact Sets** property manager opens as shown in Fig. E7-3.

- In the **Options** dialogue box, select ⊙ **Touching faces** (if not already selected).

- In the **Components** dialogue box, the **Select Components or Bodies** field is highlighted (light blue). Move the cursor into the graphics screen and click to select *both* the **Column** and the **Plate**. Select either part in any order. Names of the selected parts, **Plate 7-1-1@Assem2** and **Column 7-1-1@Assem2** are listed in the active field; see Fig. E7-3.

- Next click the **[Find Faces]** button and the software automatically identifies *any* contacting faces on these two parts. Within the **Results** dialogue box, names of the contacting faces are listed as **Contact Set-1 (-Plate 7-1-1, Column 7-1-1-)**.

- Also within the **Results** dialogue box, open the pull-down menu adjacent to **Type:** and select **No Penetration** to define the type of contact between the column and plate.

- Adjacent to **Options**, select **Node to node** as the analysis definition.

- Finally, at the bottom of the **Results** dialogue box, click the **[Create contact sets]** button.

- Click **[OK]** ✓ to close the **Find Contact Sets** property manager.

Figure E7-3 – The **Find Contact Sets** property manager facilitates definition of contact between mating parts.

Although a number of selections were involved in defining contact between the column and plate, use of the single **Find Contact Sets** property manager is much simpler than creating an exploded view; outlining contact areas using *Split Lines;* and then defining contact using the **Contact Gaps** icon as yet another step. At this point, complete the definition of the finite element model and the analysis outlined below.

    b. Create an isometric or trimetric plot showing all restraints and loads, including bolt connectors, applied to the model. This image is the finite element equivalent of a free-body-diagram. Do not show stresses on this plot. This plot should fill at least half of an 8.5"x11" page. [See also item (c) below.] Manually label bolt numbers 1 through 4 on this figure.

    c. Due to loading on this model, shear forces create the primary loads on each bolt. Using COSMOSWorks, determine the shear loads at each bolt location. Then, plot a front view of the model and on it manually sketch each shear force component and the resultant shear force at each bolt. Include labels for the magnitude of each shear force vector at each bolt.

    d. Create a plot of contact pressure between the column and plate. Exaggerate the scale of contact pressure to best depict pressure distribution on the mating parts.

    e. Manually calculate the shear force(s) at each bolt. Compare manually calculated results with finite element results and compute the percent difference between resultant shear forces at each bolt using equation [1]. Cut-and-paste two **Pin/Bolt/Bearing Force** tables from the finite element solution onto a page accompanying the manual solution. Be sure to correlate these results with the corresponding manual calculations of bolt shear force.

$$\% \text{ difference} = \frac{(\text{FEA result - classical result})}{\text{FEA result}} * 100 = \qquad [1]$$

    f. Question: Do directions of shear force components appearing in the **Pin/Bolt/Bearing Force** table correspond to forces acting *on* the bolt, or do they represent directions of shear forces exerted by the bolt on the members? Justify your answer in terms of sketches made in part (c).

2. The **Ductile Iron** "stub-end" on a pressurized tank is sealed by an **Alloy Steel** cylinder head as shown in Fig. E7-4. The cylinder head is held in place by eight **5/8 in-11 UNC, SAE Grade 8.2** steel bolts for which the modulus of elasticity, **E = 30e6 psi**, and Poisson's ratio, $\mu = 0.30$. The pipe-end is subject to an internal static pressure of 900 psi. A confined gasket seal (not shown) is located in the flange groove. Because the gasket is located in a confined groove rather than "sandwiched" between the cylinder head and flange, its low stiffness relative to other components can be ignored. Figure E 7-5 shows the basic geometry of parts in this assembly. Open files **Cylinder Head 7-2** and **Pipe End & Flange 7-2** and perform a finite element analysis to determine items requested below.

Figure E 7-4 – Bolted joint closing a "stub-end" on a pressurized tank.

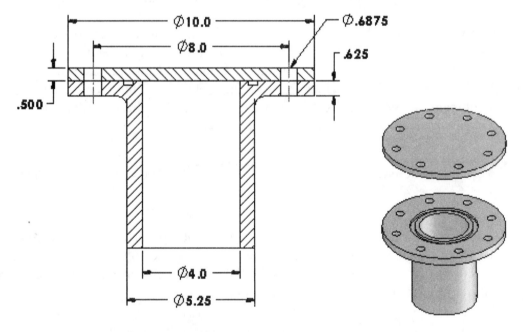

Figure E 7-5 – Basic geometry of the cylinder head and flanged pipe end. Assume the pipe segment is twice as long as shown above.

- Material: **Alloy Steel** (cylinder head)
  Ductile Iron (pipe end and flange)
  E = 30e6 psi, and Poisson's ratio $v = 0.30$ (bolt material)

- Mesh: **High Quality** tetrahedral elements

- Restraint: as necessary on the cut (bottom) pipe-end

    Contact/Gaps between cylinder head and pipe flange. See **Solution Guidance** in Exercise 1, above, for a simplified method of defining contact between the cylinder head and pipe flange.

    Bolt connector with **Nut**: Assume the outside diameter under the bolt head and nut is 15/16 in (0.9375 in).

- Pressure: **900 psi** on all internal surfaces

- Assumptions: Bolts fit in clearance holes where $D_{Hole} = 0.6875$ in
  Bolts are field tightened to a torque $T = 2540$ lb*in
  Bolt torque coefficient $K = 0.20$

Determine the following:
a. Begin by creating an assembly of the **Cylinder Head** and **Pipe End & Flange**. Use the simplified **Find Contact Sets** procedure outlined in the **Solution Guidance** section of the previous problem as you develop a finite element model that includes: material specification, restraints, load(s), bolt connectors, mesh, and solution.

b. Create a plot showing all restraints and loads applied to the model. This image is the finite element equivalent of a free-body-diagram. Do not show stresses on this plot. This plot should fill at least half of an 8"x11" page. On this plot, write a brief statement that clearly identifies the type of restraint(s) applied to the cut-end of the pipe segment shown in Fig. E 7-5.

c. Create a plot of contact pressure between the cylinder head and pipe flange. Exaggerate the scale of contact pressure to best depict pressure distribution on the mating parts. What can be stated about the uniformity of contact pressure between the mating surfaces (i.e., between the cylinder head and pipe flange)? In you answer pay particular attention to pressure distribution in spaces *between* bolts on the bolt circle that join these two parts together.

d. Due to loading on this model, each bolt is subject to its initial pre-load plus a portion of the load due to internal pressure. Use COSMOSWorks to determine the load in a representative number of bolts (at least four) by accessing results in the **Pin/Bolt /Bearing Force** window. Cut-and-paste screen images of four **Pin/Bolt/Bearing Force** windows onto the upper-half of an 8"x11" page. See also item (e) below.

e. Develop a set of manual calculations to determine the resultant tensile load in a typical bolt. Place these calculations on the lower-half of the page created in item Compare manually calculated results with finite element results and compute the percent difference between resultant bolt tensile loads using equation [1].

$$\% \text{ difference} = \frac{(\text{FEA result - classical result})}{\text{FEA result}} * 100 = \quad\quad [1]$$

Textbook Problems

In addition to the above exercise, it is highly recommended that additional problems involving bolted connections be worked from a design of machine elements or a structural analysis textbook. Bolted assemblies can be loaded by axial loads, bending loads, or shear loads, or a combination of all three. Textbook problems provide a great way to discover errors made in formulating a finite element analysis because they typically are well defined problems for which the solution is known. Typical textbook problems, if well defined in advance, make an excellent source of solutions for comparison.

INDEX

A
Animating Results, 6-22, 7-19
Assembling parts, 6-6
Automatic results plot, 4-5, 6-10
Axis, reference axis, 5-18

B
Bonded, 5-12
Bolt
   pin / bolt force table
   preload, 7-8
   torque, 7-10
Bolt Joint Example, 7-1

C
Cam Follower Example, 1-1
Circumferential Stress, 5-19
Clipping
   iso clipping, 6-20
   section clipping, 4-30 & fwd.
Connector
   bolt, 7-4
   pin / bolt force, 7-13
Contact / Gap
   bonded, 5-12
   contact set, 5-12
   define contact set, 5-11, 7-11
   shrink fit, 5-11
   no penetration, 6-15, 7-11
Contact Analysis Example, 6-1
Contact Pressure
   between surfaces, 6-24
   due to bolts, 7-21
Contact Set, (see Contact/Gap)
Convergence Check, 3-29
Copy
   copy a plot, 3-11, 5-18, 5-20
   copy a study, 3-16
COSMOSWorks
   user interface, I-11

Curved Beam Example, 2-1

Cylindrical Coordinate System
   defined, 5-17
   displayed, 5-19
Cylindrical Coordinate System, 5-19

D
Defeaturing, I-6, 3-3, 4-16, 5-5
Degrees of Freedom, I-2
   definition, I-3
   shell element, I-5
Design Check
   property manager, 2-18, 2-19
Dialogue boxes, I-12
Displacement Plot, 3-13
Draft Quality Elements, I-3

E
Elements, overview I-1 to I-6
Elements, Shell
   create, 4-12
   criteria for use, 4-3
   definition, I-3
   draft quality, I-4, I-5
   flip elements, 4-12
   high quality, I-4, I-5
Elements, Solid
   definition tetrahedral, I-3
   draft quality, I-3
   first order, I-3
   high quality, I-4
   second order, I-4
Error / Warning, 3-17
Exploded View
   bolt restraints, 7-4
   contact/gap, 6-15
   force fit, 5-10
   list selected, 5-24

F

Factor of Safety
  introduced, 2-16
Factor of Safety (continued)
  design check, 2-17
  establish criterion, 2-18
Files
  close without save, 1-24, 4-14
  open new file, 1-3, 2-2, 3-2, 4-4,
    4-15, 5-2
  save, 1-24, 4-32
Force Fit, (see Shrink Fit)
Friction, 5-12

H

High Quality Elements, I-4

I

Interference
  detection, 5-2
Interference Fit, (see Shrink Fit)
Interference Fit Example, 5-1
ISO Clipping (See Clipping)

L

List Selected
  average stress value, 5-24
  displacement, 5-26
Load
  apply to split line, 1-8, 1-10
  normal force, 3-5, 7-4
  pressure, 4-11, 4-19
  specify direction, 1-9, 1-10,
    2-10, 6-16

M

Material Properties
  from library, 1-4, 2-4, 3-4,
    4-7, 4-17, 5-4, 6-11
  custom definition, 7-8
Mates, 6-7
Mesh, I-5
  adjust mesh size, 3-17, 3-20, 3-21
  details, 1-12, 2-12
  draft quality, I-3, I-4, 3-7
  hide, show mesh, 1-12

  high quality, I-4, 3-17
  property manager, 1-11, 2-12
  mesh control, 3-22
  mesh size, 3-15, 3-20, 3-21, 4-21
  ratio, layers, 3-23
Meshing the Model
  shell element, 4-12
  solid elements, 1-11, 2-11, 3-6, 4-20,
    6-19, 7-1
Mid-side node, I-4

N

Nodes, I-1

O

Options window, 4-4, 4-24

P

Pin/Bolt Force, 7-13
Plot
  automatic results, 4-5, 6-10
  copy, 3-11, 5-18, 5-20
  multiple viewports, 3-26
Preload, 7-8
Pressure, 4-14, 4-24
Probe, 1-18, 1-22, 4-17, 4-28
Property Manager, I-12

R

Radial Stress, 5-20
Reaction Forces, 2-21
Reference Geometry, 2-7
  offset distance, 2-8
  plane, 2-7
Report
  Report generation, 5-28
Restraint
  application, 1-6, 2-5, 3-5
  force application 1-8
  property manager, 1-7, 2-5
  symbols, 4-10
  symbols, hide, 4-23
Restraint Type
  bolt connector with nut, 7-6
  immovable, 1-7, 2-5, 3-5, 4-10,
    4-18, 6-14

fixed, 1-8, 4-10
rigid body motion, 4-18, 5-8
symmetry, 4-8, 4-10, 4-17, 5-7, 6-15

S
Safety Factor, 2-16, 2-17
Second-order Elements, I-4
Section Clipping (See Clipping)
Shell Elements 4-10 to 4-19
Shrink Fits
   defined, 5-11
   contact / gap, 5-11
SolidWorks
   user interface, I-7
Solution
   introduced, 1-13
Split Line
   definition, 2-7
   creating split lines, 2-7, 6-4
St. Venant's Principle, 1-22, 4-3
Stiffness Method, I-1
Stress
   adjust display parameters, 4-26
   components X, Y, 1-16, 1-17, 3-12
   principal stress, 4-13, 4-27
   safety factor, 2-16, 2-17
   set stress limit, 2-19
   von Mises, defined, 2-16
   von Mises, introduction, 2-14
Stress Concentration Example, 3-1
Stress Plot
   add title and name, 1-17, 3-8
   boundary options, 3-9
   create a copy, 3-12, 5-18, 5-20
   fringe options, 1-18, 3-9
   ISO clipping, 6-20
   property manager, 1-17
   probe option, 1-19, 1-22, 4-28
   section clipping, 4-30 & fwd.
   settings property, 3-9
   show max annotation, 3-9
Study
   defined, 1-1
   create a copy, 3-16, 7-14
   naming, 1-4, 2-3, 3-2, 4-7, 4-16, 5-3, 6-10, 7-2

property manager, 1-3
Suppress, 3-4
Symmetry
   cut model, 6-11
   recognize model symmetry, 4-11
   restraints (See also Restraint Type.)

T
Tetrahedral Element, I-3
Thick Wall Pressure Vessel, 4-21
Thin Wall Pressure Vessel, 4-1
Toolbars (COSMOSWorks)
   loads, I-12
   main menu, I-12
Toolbars (SolidWorks)
   standard view, I-9
   view, I-9
   sketch, I-10
   features, I-10

U
Unsuppress, 4-9, 5-6

V
von Mises Stress
   defined, 2-16
   introduced, 2-14

W
Warning/Error, 3-17

NOTES: